Apatite

The Great Pretender

Contributors

Michael Bainbridge, Mark Barton, Dmitriy Belakovskiy, Dudley Blauwet, Mario Chaves
Carles Curto Milà, Donald A. Dallaire, Jordi Fabre, Carl A. Francis, Jaroslav Hyršl
John M. Hughes, Rhian Jones, Robert F. Martin, Peter K.M. Megaw, Stefan Meier, Luiz Menezes
Berthold Ottens, Alfredo Petrov, John Rakovan, Phil Scalisi, Kimberly Tait and Glen Waychunas

Editors

John Rakovan and Gloria A. Staebler with Donald A. Dallaire and David W. Bunk

Illustrators

Tina Alves, Roberto Appiani, Michael Bainbridge, William W. Besse, Dudley Blauwet
Mikhail Bogomolov, Reinhard Brauns, Joe Budd, Matteo Chinellato, Chip Clark, Jordi Deusedes
Josè Arce Diaz, Kevin Downey, Jordi Fabre, Oleg Gritsenko, Hiroyuki Imai, John Jaszczak
Rhian Jones, Boris Kantor, Kiyoshi Kiikuni, Bob King, Don Komarechka, Jesse La Plante
Marcelo Lerner, Viktor Levitskiy, Robert F. Martin, Mark Mauthner, Alana McConnon
Stefan Meier, Ikram Mohammad, Geoff Notkin, Yasuichi Okazaki, Galibert Olivier
Berthold Ottens, Natalia Pekova, Viktor Ponomarenko, John Rakovan, Ivan Rumata
John Schneider, Jeff Scovil, Guido Steger, Ryoji Tanaka, Hugo Ullhofen, Mark van Laere
Glenn Waychunas, Robert Weldon and Brad Wilson

Lithographie, Ltd. • Denver

Plate 81 "Apatite" from Reinhard Brauns' (1912) *The Mineral Kingdom* (localities as they appear in Brauns)

1. & 2. Schwarzenstein, Zillerthal, Tyrol; 3. Floitenthal, Zillerthal, Tyrol; 4. & 9. Greifenstein, Ehrenfriedersdorf, Saxony
5. Gouverneur, St. Lawrence Co., New York, U.S.A.; 6. Renfrew, Ontario, Canada; 7. Canada; 8. South Burgess, Lanark Co., Ontario, Canada
10. Snarum, Norway; 11. Kragerö, Norway; 12. Eganville, Renfrew Co., Ontario, Canada; 13. (Phosphorite) Staffel, Limburg, Hesse-Nassau (Hessen)

Contents

Apatite–A Large and Illustrious Family

John Rakovan and John M. Hughes

What's in a name? In the case of apatite, a lot. By far the most common phosphate mineral on earth, apatite is a ubiquitous minor constituent of igneous, metamorphic and sedimentary rocks. It is also found in large crystals and in large quantities in certain environments. Apatite thus must have been known for much of human history, but it was only recognized as a distinct mineral in the late eighteenth century. An ancient example of the species comes from the Neolithic archeological site of Çatalhöyük in south-central Turkey. Mellaart (1964) reports finding beads of blue and green apatite at this and other sites in Anatolia that date back to as early as the sixth millennium BCE.

The name "apatite" was first used to describe calcium phosphates, which were commonly confused for other minerals; hence, the name. "Apatite" has its origins in the Greek word απατάαο, which means "to deceive." The first substantial description of apatite as a new mineral is credited to German geologist Abraham Gottlob Werner (Werner 1788). Werner's apatite samples originated in the tin mine in the town of **Ehrenfriedersdorf** in the Erzgebirge ("Ore Mountains") of southern Germany. The tin deposit in Ehrenfriedersdorf is the putative type locality for apatite (see page 18).

Later, in the mid-nineteenth century, it was found that the chemistry of this mineral could vary among samples, especially with respect to fluorine (F), chlorine (Cl), and hydroxyl anions (OH), and the name "apatite" came to describe a mineral group in which three mineral species share the same root name (Damour 1856; Rammelsberg 1860): fluorapatite, hydroxylapatite, and chlorapatite. Their ideal compositions are $Ca_5(PO_4)_3F$, $Ca_5(PO_4)_3OH$, and $Ca_5(PO_4)_3Cl$, respectively. But a given apatite specimen may contain more than one of the elements fluorine and chlorine as well as the hydroxyl polyanion; for instance, the specific composition may be $Ca_5(PO_4)_3(F_{0.6},OH_{0.3},Cl_{0.1})$; the general formula for apatite is thus $Ca_5(PO_4)_3(F,OH,Cl)$. Because the chemical composition cannot be determined by visual inspection, these minerals are quite often referred to simply as "apatite" (*sensu lato*). The term "apatite" will be used throughout this volume when the specific composition is uncertain. In addition to the elements that define the apatites, numerous trace and minor elemental substitutions are responsible for many of the colors and luminescent properties of fluorapatite.

STRUCTURE

As with all mineral groups, the structures of these species are essentially the same and they vary only in chemistry. In other words the atomic arrangements are isostructural, and thus apatites are isomorphs of one another. The apatite structure

Apatite Group Minerals (left to right)
Mimetite; San Francisco Mine, Magdalena, Sonora, Mexico
Pyromorphite; Les Farges Mine, Correze, France
Fluorapatite, 3 cm tall; Golconda Mine, Minas Gerais, Brazil
Fluorapatite in Calcite; Liscombe Property, Wilberforce, Ontario, Canada
Vanadinite; Mibladen, Morocco
Dave Bunk Minerals specimens (ex Wendy and Frank Melanson)
Jesse La Plante photo

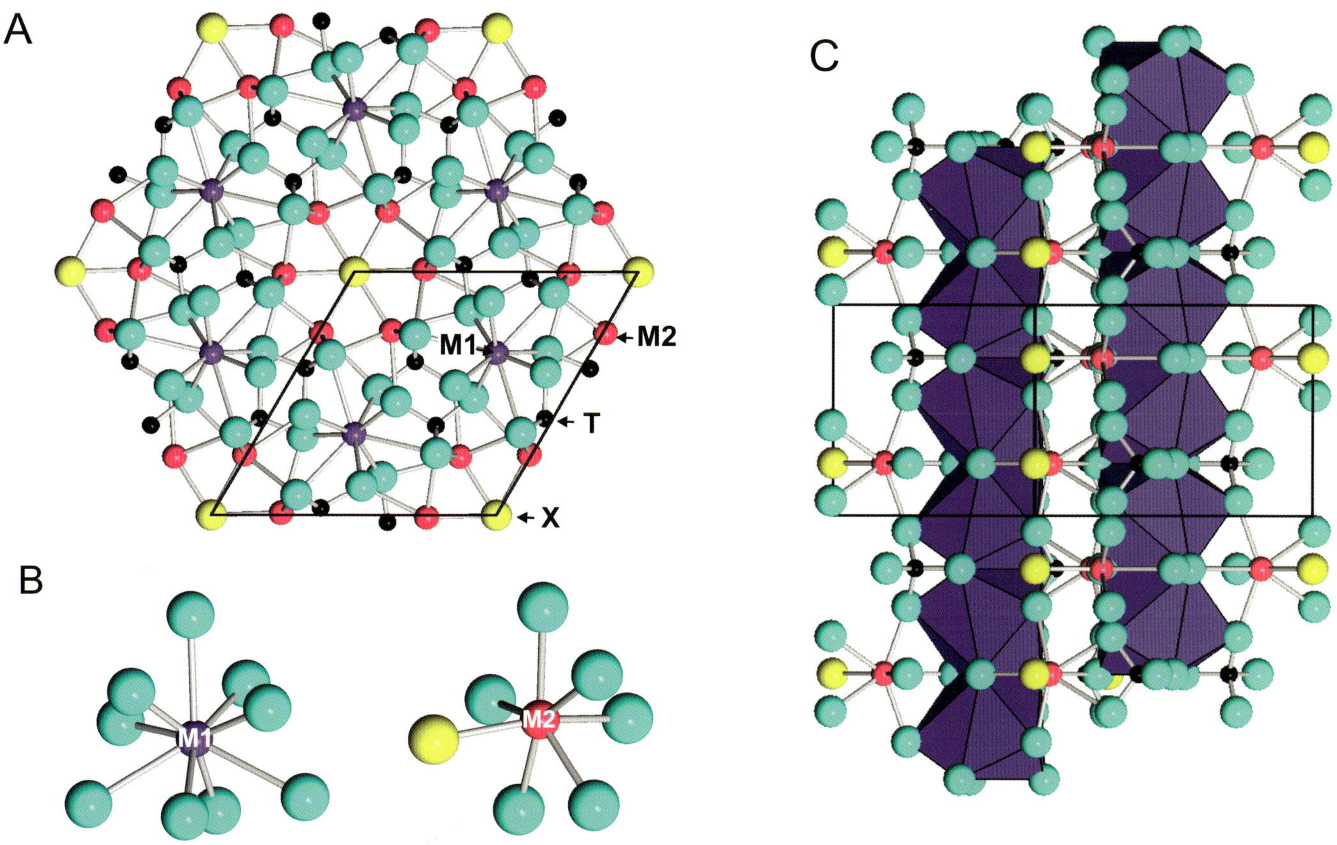

Schematics of the Apatite Structure
A) The apatite structure as viewed along the *c*-axis; several of the bonds to the M2 and T sites are obscured from this angle; **B)** Two distinct Ca sites: M1 and M2; **C)** The apatite structure as viewed perpendicular to the *c*-axis; the M1 sites are shown as purple polyhedra. The unit cell is indicated by the solid black line.
John Rakovan graphics

accommodates the inclusion of constituents other than calcium (Ca) and phosphorus (P); in fact, more than half of the periodic table of elements are known to substitute into the apatite atomic arrangement in concentrations from parts per million to tens of weight percent. Like tourmaline and the amphiboles, apatite is often referred to as a "garbage can" mineral—you can throw most anything into it. If another element, such as lead (Pb) predominates or completely substitutes for one of the essential constituents (in this case calcium) of apatite, we have a new mineral species. Pyromorphite has the apatite structure but has the ideal chemistry $Pb_5(PO_4)_3Cl$. Because of the robust nature of the atomic arrangement, more than 40 known mineral species have the apatite structure and are thus members of the apatite supergroup. In many branches of science, the term "apatite" refers to the apatite supergroup and includes any mineral or synthetic compound with the apatite structure.

The apatite structure is complex and contains three district sites on which the cations sit. The calcium and its substituents reside in the M1 and M2 sites, and the phosphorus and its substituents reside in the T-site. There are also four distinct anion sites. Three of these are always occupied by oxygen and the fourth, the X-site (commonly referred to as the column anion site), is where the fluorine, chlorine, and hydroxyl anions sit.

The exact position in the structure of the column anion site changes slightly when occupied by different ions (Hughes and Rakovan 2002). This shift makes it possible to precisely determine the relative amounts of fluorine, chlorine, and hydroxyl in a given sample by crystal structure analysis using x-ray diffraction data. The structural formula for apatite can be written as $^{[IX]}M1_2{}^{[VII]}M2_3(^{[IV]}TO_4)_3X$, where M1, M2, T and X represent the different structural sites mentioned above. The structural formula conveys information about the types of sites (i.e., coordination number) that are present and their relative abundance (i.e., stoichiometry). Roman numerals in square brackets represent the site coordination, or number of anions that are bonded to the cation at a site. Thus, calcium ions that sit on the M1 site are bonded to nine anions, all of which are oxygen. Those calcium ions that sit on M2 are bonded to only seven anions, six of which are oxygen and the seventh of which is fluorine, chlorine, or hydroxyl. The assortment of crystallographic sites with different coordinations and sizes allows for the broad chemical variation in the apatite supergroup as there are sites available for a range of ion sizes. The richly complex chemical and moderate structural variations found in the apatite supergroup are reviewed in detail in Passero et al. (2010); the mineral members are listed in the table on page 6. The focus of this volume is the calcium-phosphate apatite $Ca_5(PO_4)_3(F,OH,Cl)$. The dominant species in rocks and the vast majority of apatite specimens in mineral collections are fluorapatite.

CRYSTALS

The symmetry of most apatite crystals is hexagonal; however, positional ordering of chlorine atoms or hydroxyls in end-member chlorapatite and hydroxylapatite can lead to monoclinic symmetry (Hughes and Rakovan 2002), such as those from Bob's Lake Chlorapatite Occurrence, Frontenac County, Ontario, Canada (see page 37–39). To indicate the monoclinic symmetry these species are given the same name but the suffix "-M" is added. The lower order of monoclinic symmetry is never apparent in the external morphology, however. Observing samples by polarized light microscopy is the best way to distinguish symmetry in apatite. Apatite is most often found as euhedral to subhedral hexagonal crystals. The habit is usually prismatic—elongated along the *c*-axis—or is tabular, although equant crystals are not uncommon. The most common forms exhibited by crystals are the hexagonal prism and the basal pinacoid. Dipyramidal forms are also frequently observed, sometimes as the only form on the crystal. Some crystals, especially those found in Alpine cleft-type environments, can be quite complex, exhibiting more than half a dozen forms. Goldschmidt (1913) provides 13 pages of apatite crystal drawings and illustrates more than 17 distinct forms. Palache et al. (1951) list 30 forms that had been observed on apatite by that date.

Aggregates of macroscopic crystals showing regular habits

Apatite commonly forms prismatic crystals, as in the specimen pictured here, but it also forms tabular and equant crystals, all of which can exhibit a variety of morphologies.

Fluorapatite, 5.2 cm high
Imilchil, Meknès-Tafilalet, High Atlas Mountains, Morocco
Spirifer Minerals specimen, Jeff Scovil photo

The Apatite Supergroup

The Species and Their Respective Formulae

Apatite Group

Fluorapatite • $Ca_5(PO_4)_3F$
Chlorapatite • $Ca_5(PO_4)_3Cl$
Chlorapatite-M • $Ca_5(PO_4)_3Cl$
Hydroxylapatite • $Ca_5(PO_4)_3OH$
Hydroxylapatite-M • $Ca_5(PO_4)_3OH$
Svabite • $Ca_5(AsO_4)_3F$
Turneaureite • $Ca_5(AsO_4)_3Cl$
Johnbaumite • $Ca_5(AsO_4)_3OH$
Johnbaumite-M • $Ca_5(AsO_4)_3OH$
Stronadelphite • $Sr_5(PO_4)_3F$
Pyromorphite • $Pb_5(PO_4)_3Cl$
Mimetite • $Pb_5(AsO_4)_3Cl$
Mimetite-M • $Pb_5(AsO_4)_3Cl$
Alforsite • $Ba_5(PO_4)_3Cl$
Vanadinite • $Pb_5(VO_4)_3Cl$

Hedyphane Group

Hedyphane • $Ca_2Pb_3(AsO_4)_3Cl$
"Hydroxylhedyphane" • $Ca_2Pb_3(AsO_4)_3OH$
Phosphohedyphane • $Ca_2Pb_3(PO_4)_3Cl$
Fluorphosphohedyphane • $Ca_2Pb_3(PO_4)_3F$
"Hydroxylphosphohedyphane" • $Ca_2Pb_3(PO_4)_3OH$
Unnamed • $Ca_2Sr_3(PO_4)_3F$
Morelandite • $Ca_2Ba_3(AsO_4)_3Cl$
Unnamed • $Mn_2Ca_3(PO_4)_3Cl$
Cesanite • $Ca_2Na_3(SO_4)_3OH$
Caracolite • $Na_2(Pb_2Na)(SO_4)_3Cl$
Aiolosite • $Na_2(Na_2Bi)(SO_4)_3Cl$

Belovite Group

Fluorcaphite • $SrCaCa_3(PO_4)_3F$
Fluorstrophite • $SrCaSr_3(PO_4)_3F$
Deloneite
$(Na_{.5}REE_{.25}Ca_{.25})(Ca_{.75}REE_{.25})Sr_{1.5}(CaNa_{.25}REE_{.25})(PO_4)_3F_{.5}(OH)_{.5}$
Belovite-(Ce) • $NaCeSr_3(PO_4)_3F$
Belovite-(La) • $NaLaSr_3(PO_4)_3F$
Kuannersuite-(Ce) • $NaCeBa_3(PO_4)_3F_{0.5}Cl_{0.5}$

Britholite Group

Britholite-(Ce) • $(Ce,Ca)_5(SiO_4)_3OH$
Britholite-(Y) • $(Y,Ca)_5(SiO_4)_3OH$
Fluorbritholite-(Ce) • $(Ce,Ca)_5(SiO_4)_3F$
Fluorbritholite-(Y) • $(Y,Ca)_5(SiO_4)_3F$
Fluorcalciobritholite • $(Ca,REE)_5(SiO_4,PO_4)_3F$
Tritomite-(Ce) • $Ce_5(SiO_4,BO_4)_3(OH,O)$
Tritomite-(Y) • $Y_5(SiO_4,BO_4)_3(O,OH,F)$

Ellestadite Group

Hydroxylellestadite • $Ca_5(SiO_4)_{1.5}(SO_4)_{1.5}OH$
Fluorellestadite • $Ca_5(SiO_4)_{1.5}(SO_4)_{1.5}F$
Mattheddleite • $Pb_5(SiO_4)_{1.5}(SO_4)_{1.5}Cl$
"Hydroxylmattheddleite" • $Pb_5(SiO_4)_{1.5}(SO_4)_{1.5}OH$

Hydroxylapatite

In contrast to collector specimens, biological apatites are almost exclusively hydroxylapatite, with other notable substituents, especially carbonate (CO_3). Biological apatite crystals, particularly those that comprise bone and teeth, are extremely small with dimensions of tens to hundreds of nanometers. Because of this, scientists studying the structure and other aspects of bone apatite, by methods that require larger single crystals, use geologic specimens of hydroxylapatite as analogs (Elliott 1994). Historically, the "gold" standard for geologic hydroxylapatite has been from **Holly Springs**, Cherokee County, Georgia, in the United States (Kay, Young, and Posner 1964). Although the aesthetics of Holly Springs hydroxylapatite specimens pale in comparison to those of fluorapatite, they are superlative in size, structural quality, and chemistry for the species. No other geologic apatite has played as significant a role in our understanding of biological apatite and hence in medicine and the improvement of human health.

Holly Springs apatite is found in talc veins that cut through a small hydrothermally altered (serpentinized) ultramafic rock body of the Piedmont physiographic province in North Georgia (Vincent, McConnell, and Perley 1990). Locally, workings at this site are known as the **Verde Antique Quarry**. The crystals can reach several centimeters in size but are typically subhedral to anhedral in shape. Morphologically good crystals are rare and when found are simple hexagonal prisms terminated by the basal pinacoid. The color is uniformly a green-yellow and crystals can be quite gemmy in sections. The halogen site in the structure of Holly Springs apatite is almost completely occupied by the hydroxyl molecule making these samples near end-member hydroxylapatite.

Abundant research material once existed in museums around the country, but the demand has caused dwindling stores and the amount of extant Holly Springs hydroxylapatite may not be enough to meet future research needs. This potential dilemma has been somewhat alleviated by the recent discovery of a large quantity of hydroxylapatite at the **Sapo Mine**, Minas Gerais Brazil (Menezes 2009) (see page 54). The chemistry of these samples varies but the halogen site of the structure can be as much as 75 percent occupied by hydroxyl and only 25 percent by fluorine. While not as chemically pure as Holly Springs material, Sapo Mine hydroxylapatite may be a suitable biologic analog for many potential studies. From a collector's standpoint Sapo hydroxylapatite is arguably the finest for the species, with large aesthetic crystals often found on a microcline or quartz matrix.

Samples are found in cavities in the Sapo granite pegmatite. Since 2005 numerous pockets containing apatite have been discovered, and most samples that have been analyzed appear to be fluorapatite. The exact species cannot be distinguished without chemical or structural analysis but the hydroxylapatite samples from the most significant pockets yet discovered (in June of 2005; Menezes 2009) can be easily recognized. They exhibit flattened dipyramidal crystals with blue-green and yellow color zoning that correlates with the hydroxyl and fluorine contents: the yellow cores of these crystals are fluorapatite while the later stages of growth, indicated by the blue-green color are hydroxylapatite.

Synthetic hydroxylapatite is also an important analog for biological apatite. It is used extensively in biomedicine (Dorozhkin 2008). In fact, the most active area of apatite research today is in biomedicine, and rapid advances in the science are quickly improving implementation in humans. However, growth of large single crystals of synthetic hydroxylapatite is difficult and geologic specimens will continue to be important for scientific studies.

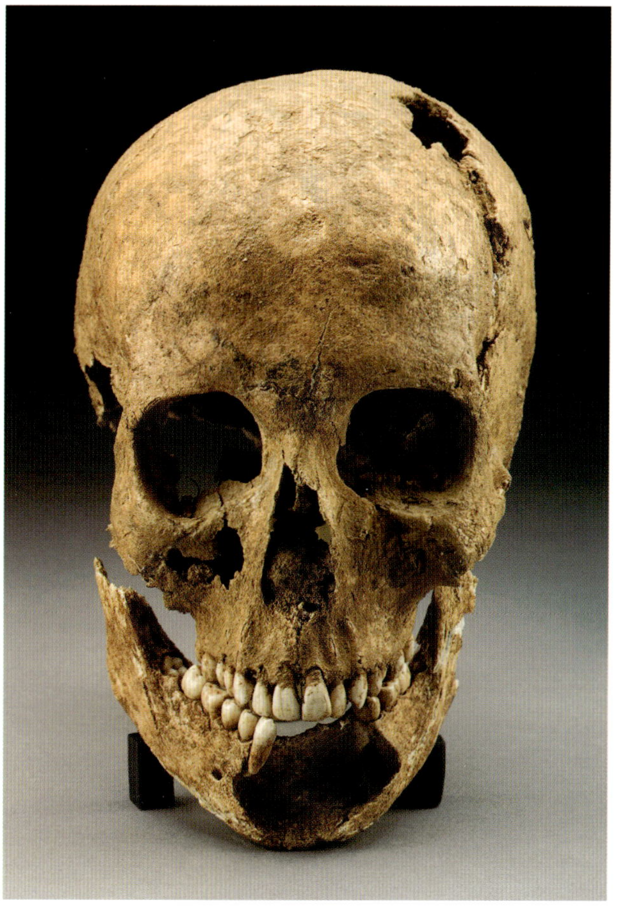

Skull of a European male, 14 to 15 years old, dating to about 1607 and found at the James Fort archeological site in Jamestown, Virginia, USA
Chip Clark photo, courtesy of the Smithsonian Institution

Apatite At A Glance

Composition
The ideal formula for apatite is $Ca_5(PO_4)_3(F,OH,Cl)$ with Z=2. Trace amounts of various elements and color centers are responsible for the various colors.

Structure
The structural formula is $^{IX}M1_4{}^{VII}M2_6(^{IV}TO_4)_6X_2$

Crystal Class
Apatite is hexagonal; its space group is $P6_3/m$ (rarely monoclinic $P2_1/b$); point group 6/m.

Unit Cell
The unit cell formula is $Ca_{10}(PO_4)_6(F,OH,Cl)_2$. The unit cell dimensions are a=9.3925(1)Å, b=9.3925(1)Å, c=6.8839(2)Å. Fluorapatite from Cerro de Mercado, Durango, Mexico, is the standard.

Hardness
Apatite is one of the standards on the Mohs hardness scale, exhibiting a hardness of 5.

Diaphaneity
Transparent to opaque

Luster
Vitreous to Subvitreous

Tenacity/Cleavage/Fracture
Brittle—sensitive to thermal and vibrational shock; poor cleavage on (001); conchoidal to uneven fracture

Density
3.1 to 3.25 g/cm^3

Color
Pure apatite is colorless to white but is colored by various factors and as a result is found in many hues and shades of pink, red, yellow, brown, green, blue, violet, purple.

Polychroism/Pleochroism
Both phenomena are weak to strong; exceptional examples include blue/yellow-green crystals from Merelani Hills, Tanzania; Slyudyanka, Russia; Mogok, Myanmar.

Color Change
Some samples show distinct color change under different light sources (see page 10)

Color Enhancement
The color of blue-green apatites from several localities in both Madagascar and Brazil is artificially enhanced to "neon" blue by heat-treating.

Luminescence
Fluorescent and cathodoluminescent at variable strengths in white, yellow, pink, blue, orange, violet

Refractive Indices
n_ω=1.631–1.650 and n_ε=1.633–1.646

Birefringence
The maximum birefringence is δ=0.002.

Type Locality
Tin deposit on Sauberg Mountain, Ehrenfriedersdorf, Saxony, Germany (see page 18)

Left: Fluorapatite "faden," 5.6 cm tall
Alchori, Shigar Valley, Skārdu District, Gilgit-Baltistān, Pakistan
John Rakovan collection, Jeff Scovil photo

Upper left: Fluorapatite, 2.8 mm
Mt. Fibbia, Ticino (Tessin) Canton, Switzerland
Domenico Preite collection
Matteo Chinellato photo

Upper right: Fluorapatite, 3.5 mm radial group
Opal Hill Mine, Riverside County, California, USA
Domenico Preite collection, Chinellato photo

Right: Fluorapatite, 1.6 cm long "spray"
Riesenwald Quarry, Baden-Württemberg, Germany
Gregor Markl collection, Jeff Scovil photo

Growth Hillocks

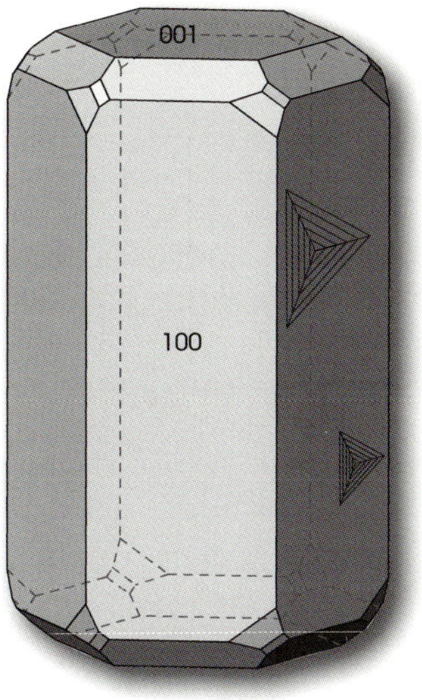

Drawings: Tabular (**below**) and elongate (**left**) crystals, two common habits of apatite. In both, the *c*-axes are oriented vertically and the $a_2(=b)$-axes run left to right. The orientations of triangular growth hillocks on the prism faces are shown.

Below: Cathodoluminescence image of a 3-sided growth hillock on a prism (100) face of an apatite crystal from the Golconda Mine, Minas Gerais, Brazil.

John Rakovan graphics

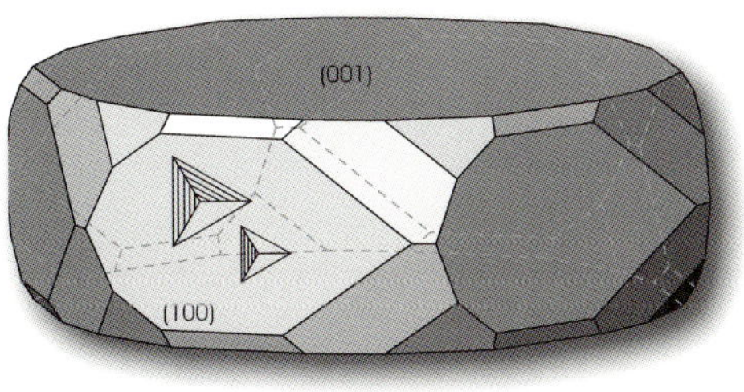

Fluorapatite, 2.59 carats
Pulsifer Quarry, Auburn, Androscoggin County, Maine, USA
GIA collection (cat. no. 134), Mark Mauthner photo

can resemble rosettes, wheat-sheafs, bowties, "faden" crystals (facing page), or scepters. A quasi-regular type of columnar aggregate resembling a "twisted herringbone," often with a hollow core (Menezes 2009), has been noted from several localities including the **Sapo Mine** in Minas Gerais, Brazil (see page 54); **Palermo #1 Pegmatite Mine** in North Groton, New Hampshire (page 28); **Pomarolli Mine**, Divino das Larangerais, Minas Gerais, Brazil; and several pegmatites in **Pakistan** (page 86). Crystals are also found in anhedral massive, granular, or globular morphologies. Sedimentary apatites are most often found as massive, cryptocrystalline crusts, concretions, stalactites, nodules, and spherical masses of radiating fibers, collectively known as "collophane."

Apatite crystals, especially those that precipitated from hydrothermal fluids, often show well-developed, growth-related surface microtopography such as striations, steps, islands, spirals, and hillocks (Rakovan 2002). Although often very subtle, these can be observed by the unaided eye, especially when crystal faces are viewed in specular reflection. As with all crystals, the orientations of growth steps are controlled by the crystal structure, and the shapes of spirals and hillocks are constrained by symmetry. Although hexagonal-shaped growth-hillocks are commonly found on the basal pinacoid faces, prism faces are most commonly striated with steps oriented parallel to the *c*-axes. Hillocks, however, are not uncommon on prism faces and if present are most often triangular in shape. Dipyramidal faces less frequently exhibit obvious growth related microtopography, but when present these features are usually more rounded than when observed on other crystal forms.

Ultimately the product of its atomic arrangement and varied chemistry, apatite has been exploited in scientific and technological applications. Although relatively soft (5 on the Mohs hardness scale) and brittle, apatite has been used as a gemstone because of the variety of colors in which it occurs. For similar reasons, rocks with abundant apatite are also used as ornamental stones.

Natural apatite is an integral component of geological and biological systems. It is the major mineral component in the bones of all vertebrates and is the fundamental source of phosphorus—an element that is critical for biological activity—on earth. Because of its geological and biological occurrence and its variety of unique properties, apatite has great significance in many areas of human interest including mineralogy, petrology, biology, medicine, geochronology, and materials science; from an anthropomorphic perspective apatite is truly a Renaissance mineral.

ACKNOWLEDGMENTS: *The authors wish to acknowledge the support of National Science Foundation grants EAR-1249459 and and EAR-0952298.*

Color and Luminescence in Apatite

John Rakovan and Glenn A. Waychunas

The causes of color and of luminescence in apatite are closely related. The highly accommodating apatite structure allows for the incorporation of many elements, often in combination, that can impart color (*chromophores*) or can lead to luminescence (*activators*). Apatite is also susceptible to other types of "defects," such as site vacancies, vacancies with trapped electrons, and point defect clusters, all of which can likewise influence color and luminescence.

COLOR

Although defect free, chemically pure apatite, $Ca_5(PO_4)_3(F,OH,Cl)$, is colorless, the photographs in this volume are vivid testament to the rainbow of possible colors found in this mineral. The causes of color in apatite are varied, and some are complex. They may involve electron transitions among orbitals or the energy states of a single substituent ion (i.e., crystal field transitions), they may be the result of simple or complex defects, or they may exhibit multiple chromophores and/or defects.

As with any transparent mineral the color that one sees in an apatite specimen may be the result of inclusions. Several examples are pictured throughout this monograph. Some crystals from the Sapo Mine, Brazil, are green due to inclusions of chlorite (see page 58). Jamesonite inclusions in apatite from Llallagua, Bolivia, make them opaque and black with a metallic reflection (see below) and inclusions of hematite in apatites from Cerro de Mercado, Mexico; Mineville, New York; and Slyudyanka, Russia (see page 108) can impart a red color to crystals.

Some of the data that have been gathered have been interpreted in different ways, leading to controversy over the exact mechanisms of color in some samples (Gilinskaya and Mashkovtsev 1995 and references therein). Furthermore, the results of color studies of synthetic apatites often do not correlate well with data from natural samples. Although some studies are fairly definitive, there is still much to learn about the details of apatite color (Bakhtin, Nizamutdinov and Vinokurov 1972; McConnel 1973).

Left: Fluorapatite, 3 cm tall
Valatscha, Graubünden, Switzerland
Canadian Museum of Nature collection (cat. no. 56492)
Michael Bainbridge photo

Middle: Fluorapatite, 6.6 cm; Dattaw Mine Mogok, Myanmar
Bill Larson collection, Robert Weldon photo

Right: Fluorapatite, 2.1 cm tall
Merelani Hills, Umba Valley, Tanzania
Dylan and Nick Stolowitz collection, Jeff Scovil photo

Facing page: Fluorapatite on albite (longwave (UVA) light) 8.6 cm tall
Bulache, Skārdu District, Gilgit-Baltistān, Pakistan
Jonathan Allred collection, Joe Budd photo

Simple Chromophores

In terms of simple chromophores, manganese is one of the most important. It can take several oxidation states and can reside on all of the cation sites within the structure. Pentavalent manganese (Mn^{5+}) substitutes for phosphorus (P) on the tetrahedral site and imparts a deep blue color (Johnson, Prener, and Kingsley 1963; Dardenne, Vivien, and Huguenin 1999; Hughes et al. 2004; Ribeiro et al. 2005). If present, even a small amount of Mn^{5+} will dominate the spectrum, and the color of apatite because of the ion's intense absorption in the visible region. Optical absorption spectra from apatite colored by Mn^{5+} show a characteristic broad absorption peak in the 640 nanometer (nm) region (graph right).

Ribeiro et al. (2005) found that heating natural blue-green apatites from Brazil changed the color to a more pure blue. They showed that the blue component was due to Mn^{5+}, but the origin of the temperature-sensitive green component was undetermined. Heat-treatment of blue-green apatites from other localities (for example skarn apatites from the Isalo area and pegmatite apatites from the Ambilobe region, both in Madagascar) also changes the samples to vibrant, sometimes neon blue (photo at the top of the page).

Divalent manganese (Mn^{2+}) can substitute for calcium in either the M1 or M2 sites (see page 4), although at lower concentrations the ion has a strong preference for the M1 site (Pan and Fleet 2002). While the role of Mn^{2+} in apatite luminescence

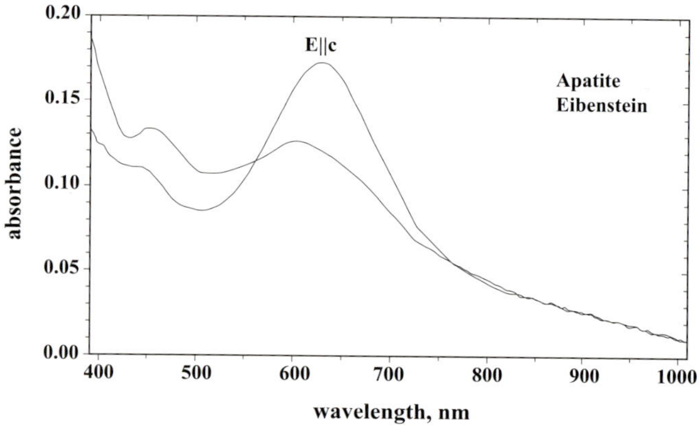

Graph: The optical absorption spectrum of Mn-rich fluorapatite from Eibenstein, Austria. Absorption is dominated by Mn^{5+} on the tetrahedral site of the structure (after Hughes et al. 2004)

Top: Fluorapatite, to 4.60 carats
South-central Madagascar
The specimens on the left are untreated, those on the right were heated to 600 degrees Celsius.
GIA Museum collection (cat. nos. 20856, 29202, 29203, 33785)
Mark Mauthner photo

Lower left: Apatite, field of view 4.5 cm wide
Alto da Cabeça Mine, Parelhas, Rio Grande do Norte, Brazil
David Bristol collection, Joe Budd photo

Lower right: Fluorapatite, 5.1 cm wide
Panasqueira Mines, Castelo Branco, Portugal
Rock Positive specimen, Jeff Scovil photo

has been extensively studied, its role in color has not. Lagerwey (1977) theorizes that rose, red-purple, and blue-purple in apatites from Panasqueira, Portugal, are the result of Mn^{3+} ions in the M1 and M2 sites. Similarities between optical absorption spectra from violet and pink apatites and silicates known to be colored by Mn^{2+} (i.e., a broad absorption peak centered around 520 nm) (Rossman 2003) suggest that Mn^{2+} may also play a role in the color of these apatites.

Defects

Interestingly, some colorless apatites will turn pink when exposed to an intense x-ray or gamma ray source. Defects in the apatite structure associated with x-ray irradiation were studied by Piper, Kravitza, and Swank (1965), who found these defects to include electrons trapped at isolated fluorine-ion vacancies (unfilled crystallographic sites that are normally occupied by ions) and fluorine-ion divacancies associated with a charge-compensating oxygen ion.

Color due to such defects is unstable and fades quickly when exposed to sunlight. Some natural pink apatites behave similarly—quickly losing color when exposed to sunlight—and thus may derive their color from similar types of defects. The intensity and richness of color in pink apatites are often highly sensitive to the source of illumination. For example, medium intensity (in sunlight) pink apatites from Chumar Bakhoor, Pakistan, are distinctly paler when viewed in blue-white LED light and are more richly colored in 3,000 Kelvin halogen light. These same apatites can be changed to a deep raspberry-red if irradiated by a gamma source (Mark Mauthner, unpublished data).

Prevailing models suggest that blue-green apatites that do not derive their color from Mn^{5+} substituting for phosphorus (as above) may instead be colored by complex color centers involving the substitution of SO_4 and SiO_4 for PO_4 and subsequent formation of SO_3^- (which causes blue) and SiO_3^- (which causes green). These groups are formed by irradiation derived from the natural decay of uranium (U) and thorium (Th) substituting for calcium in the M2 site in the crystal structure of fluorapatite (Gilinskaya and Mashkovtsev 1995; Luo et al. 2011). The SiO_3^- centers are coupled with rare earth elements on the M1 and M2 sites. Optical absorption spectra from these blue-green samples exhibit a large broad peak centered around 700–710 nm (Gilinskaya and Mashkovtsev 1995; Long et al. 2013). Color due to SO_3^- and SiO_3^- defect centers can be "bleached" by heating specimens to 400–450 degrees Celsius and 650–700 degrees Celsius, respectively.

Iron (Fe^{2+}), in trace amounts, has also been proposed as a potential chromophore in some green apatites, however, no definitive studies have been published (Knutson, Peacor and Kelly 1985; Gilinskaya and Mashkovtsev 1995). Synthetic apatite doped with chromium (Cr^{2+}) is green, but chromium has not been identified as a chromophore in natural samples (Grisafe and Hummel 1970).

Optical absorption spectra of yellow fluorapatite crystals from Cerro de Mercado, Durango, Mexico, and other localities are dominated by multiple, sharp absorption peaks, which can be attributed to a combination of rare earth elements including praseodymium and neodymium (Gilinskaya and Mashkovtsev 1995; Cantelar et al. 2001; Rossman 2003).

Lagerwey (1977) studied color in apatites from Panasqueira, Portugal, and a few from other classic localities including Ehrenfriedersdorf, Germany, and Slyudyanka, Russia. He suggests alternatives to the prevailing models, although the extant data are less supportive of his ideas. Rather than blue, Lagerwey (1977) attributes yellow-green to pentavalent manganese on the tetrahedral phosphate site in Panasqueira apatites. He also suggests that blue-green crystals from Panasqueira and blue crystals from Slyudyanka, are colored by the presence of O^- ions substituting for F on the column anion site. This is in contrast to an SO_3^- center described by Gilinskaya and Mashkovtsev (1995).

Fluorapatite on albite ("white" light), 8.6 cm tall
Bulache, Skārdu District, Gilgit-Baltistān, Pakistan
Jonathan Allred collection, Joe Budd photo

Color of apatite from some types of deposits likely reflects their chemical signatures; for example, Menezes (see page 54) points out that the color of apatite from elbaite-bearing, granitic pegmatites in Brazil are usually pink, purple, or blue, while yellow or green apatite seems to be found in aquamarine-bearing pegmatites. Further, apatite from magmatic iron deposits tend to be gemmy yellow or opaque white (and sometimes red due to inclusions of hematite).

POLYCHROIC-PLEOCHROIC-COLOR CHANGE

When discussing the color of a mineral, or any object, it is presumptively illuminated with unmodified white light. But the color of a mineral can vary, even among "white" light sources, because of variations in the spectra of different sources. The color of most apatites is barely affected by different illumination sources. There are, however, notable

Daylight

Fluorescent light

Incandescent light

Top series: Color change fluorapatite, 2 cm wide
Akjaylou, Kazakhstan
John Rakovan specimen and photos

Middle series: Dichroic fluorapatite, 2.2 cm tall
Karo Pit, Merelani, Tanzania
The color is a green-blue when viewed perpendicular to the *c*-axis (**left**) and yellow-green parallel to *c* (**right**).
Daniel Trinchillo specimen, John Rakovan photo.

Bottom series: Pleochroic fluorapatite, 2.3 cm tall
Karo Pit, Merelani, Tanzania
The image on the **left** was photographed in unpolarized light; the **middle** and **right** images in polarized light with the light vibration direction (E) parallel to the *a*-axis [100] and E parallel to the *c*-axis [001] respectively.
John Rakovan specimen, John Jaszczak photos

examples of color-change apatite. Specimens that change color from blue to violet are known from **Panasqueira**, Portugal, as is color-change fluorapatite from **Akjaylou**, Kazakhstan (Zharinov, Ponomarenko, and Pekov 2008). The latter exhibits a striking change in color from orange-red in daylight or incandescent light to yellow-gray in fluorescent light (uppermost series). It has been suggested that the color and color change seen in these samples are results of rare earth element chromophores (George Rossman unpublished data).

Even when viewed in the same light source, some minerals exhibit different colors when viewed along different crystallographic directions. Such *polychroism* can involve two (dichroism) or three (trichroism) distinct colors. Because of its symmetry, polychroic apatite is always dichroic.

Fluorapatite, colored by SO_3^- centers, shows distinct dichroism, appearing green-yellow when viewed down the *c*-axis and blue to green-blue when viewed perpendicular to *c* (middle series). Exceptional examples of this phenomenon are seen in specimens from **Slyudyanka**, Russia; **Merelani**, Tanzania; and **Mogok**, Myanmar.

Polychroism is associated with differing absorptions of light waves vibrating in different directions and is thus related to *pleochroism*. When viewed in polarized light, pleochroic minerals not only exhibit different colors from different directions, they also change color when viewed from the same direction as the vibration direction of the incident light is changed. Fluorapatite colored by SO_3^- centers also shows distinct pleochroism. When viewed perpendicular to *c* and with the light vibration direction parallel to *c*, they appear deep blue; but they appear green-yellow when viewed in the same direction but with the light vibration perpendicular to *c* (bottom series).

LUMINESCENCE

Apatite has ideal electronic properties for hosting luminescent activity. It has a large (about 8 electron volts) band gap; hence, there is good opportunity for it to host activator electronic states. It has several different cation sites (page 5), which allow substitution of a wide range of metal ion activators. In addition, it is relatively transparent to ultraviolet (UV) light; thus reasonable volumes of the mineral can be excited, leading to bright emission.

Apatite also has a versatile compositional range, with widespread solid solutions among the phosphate varieties and considerable solid solutions with the lead apatites (pyromorphite, hedyphane, mimetite, and vanadinite) as well as the other arsenate apatites (johnbaumite, turneauerite, and svabite). In many minerals ferric or ferrous iron and a number of other ions act as luminescence quenching agents, but interestingly none of the apatites incorporate significant quantities of quenching elements or ions into their structures; thus, unlike silicates in which Fe^{2+} and Fe^{3+} or Ni^{2+}, and Cu commonly quench luminescence, apatites tend to be spared this limitation.

Owing to these happy circumstances, apatite luminescence can include nearly every color in the spectrum, sometimes in combination in the same crystal (blue and gold, pink and blue, orange and yellow). Common types of apatite luminescence are frequently observed in specimens from particular geologic settings, from which the activators were derived.

Activators in the Calcium Sites

Many activators contribute to apatite luminescence and almost universally do so in concert. The major players are manganese (Mn^{2+}), lead (Pb^{2+}), europium (Eu^{2+} and Eu^{3+}), cerium (Ce^{3+}), dysprosium (Dy^{3+}), neodymium (Nd^{3+}), and samarium (Sm^{3+}) (Waychunas 2002). The luminescent colors associated with these activators are described below.

Of the two calcium sites, M1 and M2, the former is the site favored by Mn^{2+}, although the ion can be concentrated in both sites (Hughes et al. 2004; Pan and Fleet 2002). The two sites in the apatite structure have different shapes; thus the distances between the manganese and oxygen differ depending upon the site that the Mn^{2+} occupies. Consequently, the electronic structure of the Mn^{2+} is slightly different when residing on the different sites, and so too is the luminescence generated by each of the two sites: Mn^{2+} in M1 has a broad emission band centered at 587 nm (orange-yellow) while Mn^{2+} in M2 has a narrower emission band at 569 nm (yellow) (Gaft et al. 1997).

Complicating matters, major substitutions in the apatite structure can change the sizes of the sites, causing a change in the crystal fields of those sites, and thus changing the resulting emission wavelengths from ions residing on those sites. For example, Pb^{2+} and Sr^{2+} expand the sites, reduce the field, and result in emissions that shift toward yellow. When significant substitution of larger ions occurs into the M2 site, Mn^{2+} is forced to occupy the M1 site and the resulting emission is yellow. Substitution of arsenate for phosphate likewise affects the size of the calcium sites and may also slightly affect emission color, as can substitution of relatively larger Cl^- for F^-.

In addition to being a major component in the lead apatites, divalent lead (Pb^{2+}) can substitute into the calcium sites in other apatites and acts as both an activator and a sensitizer (coactivator). In other minerals, Pb^{2+} commonly produces luminescence in the near ultraviolet spectrum. This appears to be the case for caracolite, which has a strong blue-violet emission under shortwave ultraviolet light (SWUV), but a similar emission is not observed in other natural apatites. On the other hand, Pb^{2+} is a very strong sensitizer as it will markedly absorb ultraviolet radiation and transfer this energy

Identifying Activators

In general UVA (400–315 nm, roughly equivalent to longwave) only weakly excites apatite luminescence, but UVB (315–280 nm, mid-range) strongly excites many rare earths. UVC (280–100nm, shortwave) strongly excites both rare earths and Mn^{2+}, especially if Pb^{2+} is present as a sensitizer. Given this knowledge, we can identify individual activators and their sites in the apatite structure using time-resolved spectroscopic measurements.

The graphs below, showing measurements for an apatite from Norway, illustrate how activators can be separated using this technique. The idea is to excite with a very short duration laser pulse and then wait an interval (or "delay") before "gating," or turning on the detector. In A, cerium in the M1 site is preferentially observed using a delay of 50 nanoseconds because we know that the fluorescence activated by cerium decays very quickly. Applying a slightly longer delay (B) cerium is seen as an activator in both the M1 and M2 sites. If the delay is lengthened to 500 nanoseconds (C) the cerium fluorescence has disappeared, and divalent europium can be observed. After an even longer delay of 1 microsecond (D), the various trivalent REE emission bands can be seen. A milliseconds-long delay separates the Mn^{2+} emission. Using steady ultraviolet light from a typical UVB or UBC lamp, all of the emissions would occur simultaneously and be difficult to sort out.

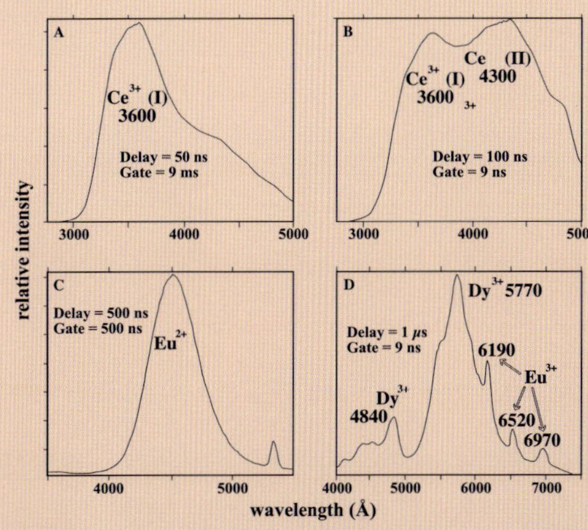

Time-resolved laser-excited spectroscopy separating activators in a Norwegian apatite (after Gaft et al. 2005)

to other activators such as Mn^{2+} or the rare earths, resulting in stronger luminescence. Antimony (Sb^{3+}) is an activator of apatite phosphors, producing strong broadband blue emission centered at about 480 nm (Blasse and Grabmaier 1994), but it has not been observed in natural apatites.

Europium (Eu^{2+} and Eu^{3+}) is an important activator in apatite. Eu^{2+}, mainly in the M2 site, produces broadband emission at about 450 nm. Eu^{3+}, like most of the trivalent rare earths, produces a set of narrow emission bands at 590, 618, 653, and 700 nm when in the M1 site and 574, 623, and 629 when in the M2 site (Gaft et al. 2005). Accordingly, Eu^{3+} may contribute to red-orange emission or to pink when combined with a blue emission from another activator.

A very strong activator of blue-violet in apatite is trivalent cerium (Ce^{3+}), a common rare earth element that probably overwhelms the effect of Eu^{2+} in most cases. Ce^{3+} in M1 emits near 360 nm and in M2 emits at about 430 nm. Most of the blue emission in apatite is likely activated by Ce^{3+}. However because of the spectral overlaps of emitting activators, the source of a given blue emission can only be verified through detailed spectroscopy including time-resolved methods.

Dysprosium (Dy^{3+}) is often mentioned as the prime activator of yellow luminescence in apatite, such as the bright emission often found in pink apatite from Afghanistan. In synthetic samples, dysprosium-activated emission is mainly from bands at 478 and 573 nm, which together would yield a green-yellow; thus, the bright yellow emission seen in natural examples must derive from more than one activator.

Trivalent samarium (Sm^{3+}) produces sharp groups of emission bands centered at about 604 (red) and 651 (orange) nm with some variation. The uranyl (UO_2^{2+}) ion may not be a structural component of natural apatite although characteristic uranyl emission is seen in sedimentary carbonate apatites. Uranyl could substitute on the phosphate site, but this is unlikely due to its positive charge. Uranium phosphates, however, often accompany apatite in pegmatites and yield green and orange (from the apatite) emissions. Uranate (U^{6+} without the uranyl complex) substitutes into the M1 site in apatite, but it causes a structural distortion and gives rise to a green emission that is different from uranyl (Rakovan et al. 2002).

Activators in the Phosphate and Halogen Sites

Pentavalent manganese (Mn^{5+}) in the tetrahedral phosphate site produces strong color in white light and gives rise to luminescence in the infrared region. Vanadate (VO_4^{3-}) is believed to produce the broadband luminescence at 580 nm in pyromorphite (Gaft et al. 2005) and may contribute to more general apatite emission if the apatite contains a vanadate component. Vanadate creates luminescence around 620 nm (orange) in synthetic phosphors. Substitution of Cl^- for F^- in the halogen site shifts emissions toward slightly longer wavelengths (more orange).

Defects, such as vacant oxygen or metal ion sites that can trap or lose an electron, respectively, also certainly contribute to apatite luminescence and thermoluminescence but neither has been well characterized.

SPECTACULAR APATITE OCCURRENCES

Apatite from **Franklin** and **Sterling Hill**, **New Jersey** (U.S.A.), may be turneaurite, $Ca_5(AsO_4)_3Cl$, or johnbaumite, $Ca_5(AsO_4)_3OH$, but fluorapatite, $Ca_5(PO_4)_3F$, is the most common. All exhibit similar luminescence, varying from a dull orange to a brighter yellow-orange. In samples in which spectroscopic studies have been done, the dominant activator is Mn^{2+}, and the subtle differences in color are likely due to small changes in the sizes of the Ca sites when arsenate substitutes for phosphate or when Pb^{2+} substitutes for calcium. Luminescent apatite makes for nice combination specimens with red-fluorescing calcite and green-fluorescing willemite.

Fluorapatite crystals from **Panasqueira**, Portugal (page 70), are known for their quality, color, and occasionally their outstanding luminescence. Most specimens are luminescent, but bright emission is uncommon. Many specimens show a pale orange to slightly yellow emission, others show a bluish emission, but some specimens show dramatic zoning (facing page, lower) in which the caps of crystals fluoresce blue and the bodies yellow. Reddish or distinctively pink emission is relatively rare. The dominant activators are Mn^{2+} (orange) and Ce^{3+} (blue-violet), and all of the rare earths contribute to create yellows, pinks, and other colors.

Most pegmatitic apatites from **New England** and elsewhere in the United States fluoresce yellow, sometimes characterized as a "chalky yellow," which is brought about by a combination of manganese and divalent ions of the rare earths samarium and dysprosium. Similar emission is seen in apatites from Pakistan and Afghanistan (photo, facing page upper). Other colors are relatively rare, but include a red-fluorescing specimen from **Greenwood**, **Maine**, that is likely a result of trivalent samarium. Blue-violet emission due to trivalent cerium is occasionally seen.

Sector-zoned fluorescent fluorapatite, Siglo XX Mine, Llallagua, Bolivia, 3 cm tall under "white" (**left**) and longwave UV light (**right**) Blue and orange luminescence is activated by different rare earth elements that are segregated between the prism and pinacoid faces and sectors. John Rakovan specimen John Rakovan photo (**right**) John Schneider photo (**left**)

Magmatic magnetite-apatite deposits are a rich source of strongly fluorescing apatites, which generally have low manganese and high REE contents. The dominant rare earths are cerium and europium; hence blue-violet or blue emissions are most common. The presence of some manganese or samarium can lead to pink emission; yellow emission is rarer. Luminescence is stronger under UVB. Examples of yellow apatites from this type of environment that fluoresce blue include crystals from the **Cerro de Mercado Mine**, Durango, Mexico; the **Anti Atlas Mountains** in Morocco; and the iron ore deposits of southern Utah.

Specimens from the **Cerro Grande** and the **Siglo XX** mines in Bolivia (see page 67) are spectacular in terms of both size and luminescence. Specimens are generally brightly fluorescent and many exhibit sector zoning in emission colors (photo facing page, lower). Manganese and lead contents must be low in these crystals, and there is only a small difference in color between UVB and UVC excitation.

LUMINESCENCE ZONES

Apatite often exhibits zoning of luminescence as well as color. These zones are most commonly either concentric (around the center of the crystal) or sectoral (under different crystal faces) (Rakovan and Waychunas 1996; Rakovan 2009). Under cathodoluminescence, in which the activators are excited by electrons rather than ultraviolet light, intrasectoral zoning may be observed (see sidebar, page 9) (Rakovan and Reeder 2004).

Luminescence and color variations, ultimately related to concentric zoning but appearing similar to sectoral zoning, can occur when the last stages of crystal growth take place preferentially on certain faces and not equally on all crystal faces; the specimen pictured lower right is an example.

PHOSPHORS AND LASERS

The luminescent properties of apatite have been the foundation of the phosphor, laser, and lighting industries for many decades (Henderson and Marsden 1972). Synthetic fluorapatite, containing small amounts of substituent manganese (Mn^{2+}) and antimony (Sb^{5+}), have been used in fluorescent light bulbs and phosphor tubes since the early 1940s. In this application, the apatite is commonly referred to as "calcium halophosphate." When energized, mercury vapor in the phosphor tube emits ultraviolet light with a large proportion of wavelength 253.7 nm. This light is then absorbed by the doped apatite, which in turn fluoresces with two broad emission bands that appear white to the eye. Small variation in the chemistry of the apatite is used to control the exact color balance, allowing for bulbs with "warm white," "white," and "daylight" emissions.

And since 1990, a new generation of fluorescent bulbs has become dominant on the world market. Known as "triphosphors," these bulbs incorporate three different luminescent materials with three individual emissions; red, blue, and green. These three combine to create a whiter light that allows more accurate representations of color in illuminated objects than do the older phosphor mixes. Rare earth elements are the activators in these new phosphor materials, and the blue-emitting phosphor is commonly a europium-doped apatite.

Above: Fluorapatite on albite under shortwave (UVC) light
8.6 cm tall
Bulache, Skārdu District, Gilgit-Baltistān, Pakistan
Jonathan Allred collection, Joe Budd photo

Below: Fluorapatite under (UVC) light, 10 cm wide, from Panasqueira, Portugal. The yellow is mainly Mn^{2+} activated and the violet is Eu^{2+} and Ce^{3+} activated. The zoning is the result of late-stage preferential growth at the ends of the crystals.
Glenn Waychunas specimen and photo

The luminescence emission from apatite can be optically amplified by stimulated emission of photons; in other words, it can act as a laser; indeed apatite has been extensively investigated as a rare earth element activated lasing material.

Color and luminescence in apatites not only make these minerals attractive to collectors they are also of scientific interest and have commercial applications.

Left: A retired *Zinnerz* ("tin ore") train car at the Ehrenfriedersdorf Tin Mine at Sauberg
Photo courtesy of Deutsche Rohstoff

Below: Abraham Gottlob Werner
Photo courtesy of Bergakademie Freiberg

Facing page, upper left: Fluorapatite
Field of view 9 cm wide
Ehrenfriedersdorf Tin Mine at Sauberg
Saxony, Germany
Bergakademie Freiberg collection
Stefan Meier photo

Facing page, lower left: Fluorapatite, 4.5 cm wide
Ehrenfriedersdorf Tin Mine at Sauberg
Saxony, Germany
Karlheinz Jankowski collection
Stefan Meier photo

Facing page, right: Fluorapatite, 5.2 cm tall
Ehrenfriedersdorf Tin Mine at Sauberg
Saxony, Germany
Annemarie Grossman collection, Jeff Scovil photo

Facing page, lower right: Ehrenfriedersdorf Tin Mine at Sauberg
Photo circa 1930, courtesy of Tonik Güttler

Type Locality–Ehrenfriedersdorf

by Stefan Meier, *Marktredwitz, Germany*

The Erzgebirge ("Ore Mountains") lie on the boundary between the German state of Saxony and the Czech Republic. The town of Ehrenfriedersdorf is nestled in these mountains and is home to Germany's earliest tin mine, which was discovered in and worked since 1240. By 1300 the near surface alluvial deposits had given way to primary tin ores, which were being accessed via underground mines, through which miners also came across rich silver ores. By the early fifteenth century, tin production had reached 150 tons per year, and Ehrenfriedersdorf was the largest mining town in the Erzgebirge (Kreibich & Güttler 2012).

Production waned in the sixteenth and seventeenth centuries, but the mines were once again hopping by the early eighteenth century, when more than 300 miners were exploiting the Sauberg Mountain deposit via 50 pits. World wars I and II brought increased demand for tin and tungsten, and output at Ehrenfriedersdorf again increased to keep pace with the need for these strategic metals. Saxony fell to the Soviets after World War II and became part of the German Democratic Republic ("East Germany") in 1949. The Soviets mined the Ehrenfriedersdorf deposit from about that time until German reunification in 1990. During those nearly four decades, 9.75 million tons of tin-tungsten ore were recovered and processed.

Early on, note was made of the numerous minerals that were found in the veins and crevices associated with the tin lodes. Among these was a species with six-sided crystals that was variously mistaken for a number of other minerals including aquamarine and quartz. Ignatz von Born (1742–1791) studied apatite from the famous topaz locality of Schneckenstein, and Freiberg Mining Councilor Johann Friedrich Wilhelm von Charpentier (1738–1805) worked on specimens from the tins mines at Ehrenfriedersdorf. Each recognized that the material was a distinct, previously undefined species in 1776 and 1778, respectively, but neither scientist gave the new mineral a name (Hintze 1933). Charpentier noted (1778) that his specimens contained green crystals similar to beryl and that comparable material had been found in quartz and fluorite pockets in the **Sauberg Pit**, **Leimgrube Vein**, **Lorenz Lode**, all on Sauberg as well as in the **Großen Vierung Ore Field**, located about 100 meters east.

German mineralogist Abraham Gottlob Werner (1749–1817) first came across apatite crystals while sifting through specimens at the Academic Mineral Cabinet at the Freiberg Mining Academy (*Bergakademie Freiberg*), where he had taken over as curator from Christian Hieronymus Lommer in 1775. Of the specimens from Ehrenfriedersdorf labeled by Lommer as *Aquamarinfluß* and *Krisolithfluß* (roughly meaning "light blue fluorite" and "green fluorite"), Werner (1788) noted,

> *Among the fluorspars were two specimens with little, short, hexagonal prismatic crystals with tapered edges, of which one has a pale mountain-green color, and the other a pale leek-green, approaching olive-green color. On both of these samples the crystals sit on a druse of*

columnar crystalline quartz, and the latter were also in and on a mixture of black sphalerite, fluorspar, and green earth.

In 1780 Werner visited the tin mines in Ehrenfriedersdorf while working in the Erzgebirge and found a few samples of the unnamed species. He chemically analyzed them and found them to contain calcium and phosphoric acid, but Werner did not detect fluorine in the specimens. Werner also ascertained physical properties including density, hardness, fracture, luster, and crystal shape.

Werner chose a name for this new mineral, which had so often deceived mineralogists: the Greek word ἀπατάω (sic) (απατεών, "apatite"), meaning "deceiver." Werner, however, did not coin the term. "Apatite" had been used to refer to the same species by Carl Abraham Gerhard (1738–1821), Professor of Mineralogy and Mining Sciences at the Berlin Mining Academy (Berliner Bergakademie), in his book *Grundrisse des Mineralsystems* (1786). Gerhard, however, had only completed melting experiments and had not done the analyses and mineralogical description necessary to conclusively establish his apatite as a new mineral species (Werner 1788). Thus, although samples from Schneckenstein were among those first thought to have been a potential new species, the specimens which were used for the first definitive analyses and description (Werner 1788)—the type samples—originated in the Sauberg mines at Ehrenfriedersdorf.

Werner's *Kurze Klassifikation und Beschreibung der verschiedenen Gebirgsarten* has been commonly cited in English literature as his first publication on apatite. This however is incorrect, as Werner does not mention the word

"apatite" in the publication, which is a description of different rock types, with very little information about individual minerals. Ehrenfriedersdorf is mentioned but in the context of granite and tin deposits not in the context of apatite. Furthermore, the year of publication is often given as 1786, but the actual year was 1787.

The type material deviates markedly from the lilac apatite specimens that were discovered much later (1900–1920) in the **Greifensteine** granite quarry, also in Ehrenfriedersdorf (Weiss 2012). Although originating in deposits in the same town, the type material reflects the characteristic paragenesis of the tin

Left: View of the columnar outcrops of granite at the Greifensteine Natural Park in Ehrenfriedersdorf, Saxony
Stefan Meier photo, 2012

Lower left: Fluorapatite
5.5 cm wide
Greifensteine Quarry,
Ehrenfriedersdorf, Saxony, Germany
Frank Spallek collection
Stefan Meier photo

Lower right: Fluorapatite, 3 cm wide
Ehrenfriedersdorf Tin Mine at Sauberg
Saxony, Germany
Tonik Güttler collection
Stefan Meier photo

The two lower photos and lower left photo on page 9 were previously published in *Lapis* and are reprinted here with the kind permission of Christian Weise Verlag.

ore veins of the Sauberg Pit rather than the that of the clefts or quartz veins of the Greifensteine granite. The Greifensteine quarry is now in a protected natural area.

Gustav Rose (1798–1873) first detected fluorine in apatite in 1827. Rose studied samples from Ehrenfriedersdorf; **Snarum**, Buskerud, Norway; **Cabo de Gata**, Andalucía, Spain; **Greiner** and **Valtigl**, Tyrol, Austria; **Arendal**, Aust-Agder, Norway; **Saint Gotthard Mountain**, Switzerland; **Chudleigh**, Devon, United Kingdom; and from the **Laacher See**, Rheinland-Pfalz, Germany. He noted (1827),

> *The apatites are now either compounds of one atom of calcium chloride with three atoms basic phosphate of lime, or one atom of calcium fluoride with as much calcium phosphate, or mixture of the two ... Let us call the former chlorine apatite, the latter fluorine apatite.*

Over time, apatite has been increasingly better defined and further differentiated, the "deceiver" is today a series of distinct mineral species, each with an associated type locality. The original sample from the Sauberg Pit was later found to be fluorapatite $Ca_5(PO_4)_3F$. The year of discovery of apatite *per se* is 1788 (Werner), with the specific fluorine-dominant (fluorapatite) and chlorine-dominant (chlorapatite) apatites attributed to Rose in 1823. The type material should not be credited to Rammelsberg, nor shold the description date be 1860, although both are commonly reported.

For generations, collectors and scientists have used the generic term "apatite" in reference to any of the species not specifically identified as fluorine-, hydroxyl- or chlorine-bearing calcium phosphate. When the specific species is known, the name "fluorapatite," "chlorapatite," and "hydroxylapatite" are used. For a short time in the mid-2000s, the International Mineralogical Association (IMA) Commission on New Minerals, Nomenclature and Classification, renamed the apatite group minerals according to their respective chemistries, apatite-(CaF), apatite-(CaCl), and apatite-(CaOH) (Burke 2008), but the traditional nomenclature was back in place by the end of the decade (Pasero et al. 2010).

ACKNOWLEDGMENTS*: For their generous support, the author would like to thank Tonik Güttler, Alexander Hölzel, Angela Kiessling, Andreas Massanek, Frank Spallek, Stefan Weiss, John Rakovan, and Gloria Staebler.*

The Geology of Apatite Occurrences

by Robert F. Martin and John Rakovan

The geologic environments in which apatite is found are numerous. Apatite forms under conditions of temperature and pressure ranging from those found in shallow sedimentary environments to those deep within the mantle. Because of the abundance of its constituent elements and its wide stability range, apatite is the most abundant phosphate mineral in the Earth's crust and mantle. It is a minor constituent in many types of igneous, metamorphic, sedimentary, and hydrothermal rocks, and in some specific occurrences, apatite can be the dominant mineral in the rock.

Coverage of all rock types and environments in which apatite is found is beyond the scope of this chapter. Here we discuss the most significant environments in which apatite specimens of large size and significant abundance are found. This leaves out phosphorite deposits, which are very large but very fine-grained. The interested reader can consult Filippelli (2011) for a recent review on of these economically important deposits. For the most part, apatite crystals of interest to the collector form either by crystallization from an igneous melt or by precipitation from a hot aqueous fluid.

IGNEOUS APATITE

Apatite is the common mineralogical expression of the small amount of phosphorus present in virtually all melts generated in the Earth's upper mantle and crust. Phosphorus contributes to a decrease in the temperature of final crystallization of the melt. As a consequence, apatite generally is not among the early products of crystallization but rather appears as a fine-grained accessory after most of the silicate melt has crystallized. Its size is a fraction of a millimeter, in general, and crystals are usually subhedral to euhedral. Its dimensions and habit very much depend on the rate of crystal growth. An acicular habit signals a high degree of supersaturation of apatite in the melt and can be expected near the outer contact of the intrusive body; more stubby prisms are expected where the rate of crystallization was slower, in the interior of the igneous complex.

The discussion of igneous apatite occurrences will focus on five distinct settings: 1) magnetite-apatite deposits formed from an ore magma, 2) cumulate apatite in large mafic complexes, 3) apatite in granitic pegmatites, 4) apatite in marble (crustal silicocarbonatites) in collision-related orogenic belts, and 5) apatite in alkaline igneous rocks.

Magmatic Magnetite-Apatite Deposits

The most famous deposit of magnetite crystallizing from an ore magma is located on the flanks of the **El Laco** volcano, in the High Andes, northern Chile. The 300-million-ton deposit of virtually pure massive magnetite is situated at an elevation of close to 5,000 meters; reaching it is a challenge. But it is fair to say that the majority of ore-deposit specialists who have made the trek have come away convinced that lava flows consisting largely (more than 98 percent) of magnetite do indeed exist.

The structural and textural evidence strongly favors the existence of an iron oxide melt that vesiculates like an uncorked bottle of champagne, with exsolution of dissolved gases to form numerous bubbles in the melt, as it rises into the flanks of a Pliocene andesitic-rhyodacitic volcano. Crystal-growth textures provide convincing evidence of a quench of the magma in response to vesiculation, combined with a drop in temperature (Henríquez and Martin 1978). The addition of phosphorus to the oxide melt lowers the temperature of its ultimate crystallization, so that fractional crystallization of magnetite can be expected to progressively enrich the melt in phosphorus. In other words, as magnetite crystallizes and settles, the remaining magma becomes more enriched in phosphorus until it is supersaturated with apatite, which then also starts to crystallize.

Fluorapatite is locally abundant in dikes, veins, and subvolcanic bodies, but it is only a minor accessory phase in the massive ores (Naslund and Henríquez 2004). Where it is unaltered by hydrothermal activity, the primary igneous apatite crystals are transparent and colorless to pale yellow. Prisms of apatite and columnar crystals of magnetite and minor augite flare outward from a growth surface (example below). Euhedral crystals up to 7 centimeters in length have been encountered in magnetite deposits; they occur with increasing abundance and size in and near cavities, which are situated toward the top of extrusive flows (Broman et al. 1999).

Fluorapatite crystals have nucleated along the contact of a vesicular magnetite-dominant flow. El Laco magnetite deposit, on the flanks of the El Laco andesitic volcano, northern Chile. Field of view 12 cm, R.F. Martin photo

Giant Apatite Crystals from the Karur-Kangayam Gemstone Belt

by Berthold Ottens

The high pressure and temperature metamorphic rocks of the **Karur-Kangayam** gemstone belt in the Indian state of Tamil Nādu are well known for producing a wide variety of gem minerals including ruby, sapphire, cordierite, and garnet (Santosh and Collins 2003; Ottens 2011). Granitic intrusions and pegmatites associated with these metamorphic rocks are the source for fine collector specimens of quartz (smoky quartz and amethyst), beryl (variety aquamarine), schorl, and fluorapatite. Most of the pegmatites show a giant crystallization structure and are mined for industrial grade feldspar and quartz. Open pockets with well developed crystals are rather rare.

Giant apatite crystal from Aravakkurichchi, Tamil Nadu, southern India
Berthold Ottens photo

When a pocket with gem-grade specimens is encountered by miners, they usually break the crystals off of the host rock and smuggle them unwrapped in bags to local gemstone dealers. In only a few cases have remarkable crystals been carefully preserved and offered on the mineral market. One such find occurred in 2009 in an open-pit feldspar mine in **Aravakkurichchi**, about 40 kilometers south of Karūr. Most work in the mine is done by men with very simple tools and without any heavy blasting thus allowing for the recovery of specimens with no major damage. The pegmatite, with an exposure of approximately 100 by 50 meters, has several times produced quartz crystals up to 1 meter long. In the summer of 2009 a remarkable pocket with a small number of exceptionally formed, giant apatite crystals—the largest weighing approximately 30 kilograms—was unearthed. The big green crystals show good transparency and well developed, lustrous faces. Their habit is dominated by the prism, pinacoid, and dipyamidal forms. Interestingly no minerals other than apatite and feldspar were found in the pocket. In previously discovered pockets, only quartz or beryl had been found with the feldspar. Specimens with other associations have not been observed.

Very rarely smaller apatite crystals labelled generically from the Karur-Kangayam gemstone belt are offered to mineral collectors. These yellow-green, transparent crystals reach 4 centimeters in diameter and a well-developed equant to prismatic habit.

Giant apatite crystals are known from other world wide localities such as **Kovdor**, Kola Peninsula (Kol'skiy Poluostrov), Murmanskaya Oblast', Russia; **Slyudyanka**, near Lake Baikal (Ozero Baykal), Irkutskaya Oblast', Russia (see page 103); and **Renfrew County**, Ontario, Canada (page 38). However, for size, luster, transparency, and morphologic perfection, the crystals from Aravakkurichchi may be the finest giant apatites yet found.

The Chilean deposits are commonly compared with the Kiruna magnetite-apatite deposit in Sweden (Nyström and Henríquez 1994) and are referred to as Kiruna-type deposits.

In all cases of magmatic magnetite-apatite deposits, there is a significant hydrothermal overprint. Many authors have been more impressed with the importance of hydrothermal activity at these sites, to the point of attributing the entire association to hydrothermal rather than magmatic processes. Thus, the genetic affiliation of these deposits remains a source of controversy; so too are the source and mechanism of formation of the ore magma itself.

Well-known specimen-producing magmatic magnetite-apatite deposits include **Cerro de Mercado**, Durango, Mexico (see page 48); the **Iron Springs district**, Iron County, Utah, and the **Pea Ridge Mine**, Sullivan, Missouri, both in the United States (see page 28).

Cumulate Apatite in AMCG Complexes

Major occurrences of igneous rock suites containing anorthosite-mangerite-charnockite-granite (AMCG), typically of Proterozoic age, are commonly associated with accumulations of titanium-rich magnetite, ilmenite, and apatite. Most petrologists agree with Emslie (1978) and McLelland et al. (2010) that this association formed from mantle-derived magmas (anorthosites) and crust-derived magmas (granites, syenites), which rose together in the crust after a major collision of continents. The association is particularly well known in the Grenville orogen of eastern Canada and the northern United States, but it also occurs in younger orogens that have experienced a major collision. Apatite is prominent, usually as an accessory phase, in cumulates formed in association with the anorthosites.

These are massive rocks, devoid of cavities. The fluorapatite attains a grain size of one centimeter in the gabbronorite variety of this deposit type, but the crystals are generally smaller and clustered together in nelsonite. Such rocks do not typically yield apatite crystals of interest to collectors, but they offer a promising raw material for the concentration of apatite, always in demand for the fertilizer industry and recently considered as an ore for the rare-earth elements.

Fluorapatite with muscovite, albite and lepidolite, 2.9 cm tall
Otjua Mine, Karbib, Namibia
Marty Zinn collection, Jeff Scovil photo

Granitic Pegmatites

Granitic pegmatites are a well-known source of apatite crystals, but not all occurrences of granitic pegmatite are created equal. Černý and Ercit (2005) distinguished two families of granitic pegmatite: one they referred to one as "LCT" because of its enrichment in lithium, cesium, and tantalum, and to the other as "NYF" because of its relative enrichment of niobium, yttrium, and fluorine. The petrogenetic processes contributing to the formation of these two types of granitic pegmatite differ radically in a way that reflects the tectonic setting at the time of emplacement (Martin and De Vito 2005). Only the LCT pegmatites are likely to be phosphorus-enriched and thus yield apatite specimens. The famous apatite-producing pegmatites including those of Maine, New Hampshire, and California in the United States; Minas Gerais in Brazil; Gilgit-Baltistan in Pakistan, and the Nuristan region of Afghanistan are all of LCT-type. Other apatite-bearing pegmatites are found worldwide, the **Otjua Mine**, Erongo Region, Namibia, for example. And a find of exceptionally large fluorapatite crystals was recently made from a pegmatite in **Aravakkurichchi**, Tamil Nadu, India (see the sidebar on the facing page).

A zoned LCT pegmatite is likely to contain a primary apatite-group mineral in the wall zone, the first part of the intruding pegmatite-forming melt to crystallize. The melt at this stage is relatively calcic and thus produces a sodic plagioclase such as oligoclase or andesine; the apatite is likely to be acicular because it grows rapidly. It is generally oriented more or less perpendicular to the external contact (e.g., Sirbescu et al. 2009). Moving from the wall zone toward the core of the pegmatite, there is commonly a progressive drop in the amount of apatite, as the melt progressively becomes depleted, not in phosphorus but in calcium.

If the felsic melt is phosphorus-enriched, phosphate nodules are in some cases present near the outer contact of the core zone. These nodules may be the expression of an immiscibility relationship between a phosphate melt and the evolved silicate magma. The primary phosphate in these nodules is generally triphylite, $LiFe^{2+}(PO_4)$, which may be manganese-bearing. As the degree of fractional crystallization progresses to the stage of core formation, the amount of calcium drops to the level of a trace constituent, such that there is very little primary apatite either in the silicate melt or in the phosphate nodules.

Yet in miarolitic pegmatites, those containing open crystal-lined cavities or vugs, a mini-flood of calcium is evident (Jahns 1982), such that fluorapatite appears in much larger, stubbier crystals, which are highly prized by collectors. The late influx of calcium is attributed to reactions taking place beyond the outer contact of the pegmatite bodies. A corrosive aqueous fluid exsolved from the pegmatite-forming melt is actually a dilute solution of hydrofluoric and boric acids. It "back-reacts" with (i.e., replaces) the previously formed primary minerals as the pegmatite "stews in its own juice." The corrosive fluid also leaks into the adjacent country rock, where it aggressively reacts with plagioclase in the country rock, releasing calcium and sodium into the fluid. This modified fluid re-enters the pegmatite, where it is able to convert microcline to albite + muscovite. The new supply of calcium also reaches the internal portions of the pegmatite. There, the fluid can form calcic tourmaline-group minerals (including liddicoatite) and apatite (Martin 2013). Fluorite and calcic accessory minerals such as pyrochlore, microlite, and fersmite are formed by the same mechanism in NYF-type pegmatites (Martin & De Vito 2005).

Landes (1925) was a pioneer in evaluating the order of crystallization of minerals in granitic pegmatites in south-central Maine, which are phosphorus-rich and representative

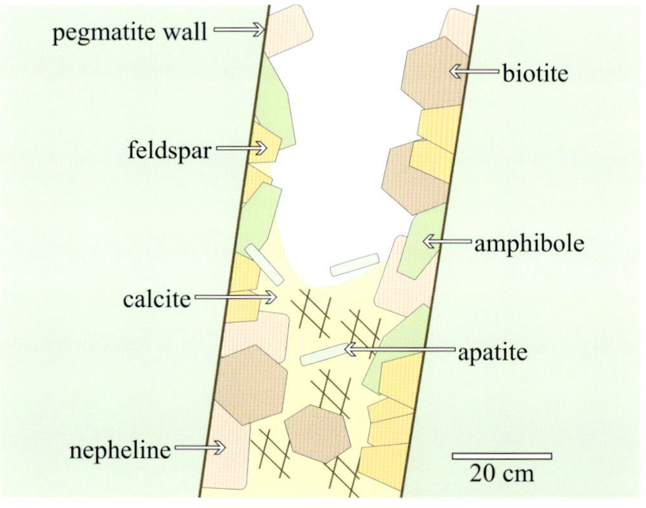

Above: A 12 cm long stack of bent fluorapatite prisms with a film of calcite between each crystal. The crystals were stacked (by settling?) in a carbonatitic melt, then subjected to mild deformation and cracking while there was still melt available to fill the tension gashes that developed. It is from Wakefield Township, Québec, Canada.
Ferrier collection, Redpath Museum, McGill University (cat. no. F4486)

Above: A schematic illustration of a calcite-cored silicocarbonatite pegmatite. Natural dissolution of the calcite has liberated euhedral crystals of apatite and various silicate minerals. William W. Besse graphic (after London 2008)

Right: Globule of orange calcite trapped in a 4-cm-wide crystal of fluorapatite, such globules are interpreted to represent melt inclusions
Yates Mine, Otter Lake Québec, Canada
Darryl B. MacFarlane collection

of the LCT-type. Landes became convinced that the pockets are secondary, created through the dissolution of primary minerals by the aggressive hydrothermal magmatic fluid, as explained above. The gem-quality blue and purple fluorapatite found in cavities in certain of Maine's pegmatites is commonly manganese-enriched; thus the phosphate, fluorine, and manganese in the fluorapatite of those pockets are contributed by the pegmatite system, whereas the calcium largely originates outside the system.

Marble in Collision-Related Orogens

The Central Metasedimentary Belt (see page 38) of the Grenville Province of southwestern Québec, southeastern Ontario, and New York State is known for occurrences of what are commonly known as "vein-dikes," a fence-sitting term (are they veins, i.e., hydrothermal, or are they dikes, i.e., magmatic?) introduced by Ellsworth (1932) and employed by Moyd (1990). Joyce (2006) reviewed the attributes of vein-dikes, and London (2008) illustrated the essential features of this unusual type of carbonatitic pegmatite (middle left). They are generally cored by orange or pink calcite, or by calcite and anhydrite, with accessory fluorite. The vein-dikes carry a wide variety of euhedral crystals that have nucleated on the walls and in some cases may be suspended, completely surrounded by the calcite (± sulfate) matrix.

The natural dissolution of the calcite matrix exposes pristine euhedral crystals of apatite, titanite, phlogopite, microcline, zircon, and nepheline, all of which are well illustrated by Joyce (2006). After sampling dozens of these vein-dikes in the Grenville Province, Moyd (1990) concluded that the nature of the euhedral crystals in the carbonate pegmatite is a function of the mineral assemblage in the host rock; thus, euhedral nepheline crystals are found only where the host rock contains nepheline, and so on for the other mineral species. His important observation points to the ability of the carbonate-sulfate-fluoride medium to dissolve and reprocess the host rock. By incorporation of silicon by assimilation of silicate host rocks, the carbonate melt develops a silicocarbonatitic composition.

Globular inclusions of melt trapped in apatite and in all of the coexisting silicates (facing page, lower) are a clear indication that this medium was indeed a melt and that the melt crystallized very quickly in the deep crust. Insight into the rate of crystallization comes from the fact that the crystals are "frozen" in the calcite matrix: as the viscosity of the melt phase was probably comparable to that of syrup, the melt phase had to freeze before the detached or unattached crystals had sunk to the bottom of the body of silicocarbonatitic magma. A quench of the melt phase probably occurred by vesiculation and escape of a CO_2 (and possibly SO_2) gas phase upon crystallization (i.e., a vapor pressure quench rather than a thermal quench). While the melt was crystallizing, there was evidently limited movement of slurries locally, such that a preferred alignment of apatite prisms ensued. A sample from Wakefield Township, Québec (facing page, upper), Canada, shows a cluster of such aligned crystals of fluorapatite, each perfectly molded by orange calcite. Note that the crystals have been bent, and orange calcite fills cracks that cross the broken crystals. The perfect molds and the tension gashes crossing the prisms and enclosing fragments of apatite attest to the mobility of the carbonate as a melt.

There is compelling evidence that the vein-dikes are in fact dikes of a newly recognized type of igneous rock, referred to as crustal silicocarbonatite (CSC) by Morteani et al. (2013). Marble in the lower part of the crust, thickened as a result of collision of continents toward the end of the Grenville orogeny, and heated by the influx of fluids and magmas issued from the rising asthenospheric mantle, actually began to melt. The same context exists in other orogenic belts in which there has been a major collision of continental plates, for example at the end of the Pan-African orogeny in eastern Africa, including Madagascar.

The isotopic composition of the carbonate matrix and of the associated silicate minerals is clearly crustal, and far from values expected of mantle-derived rocks; the values are inconsistent with a mantle-derived carbonatitic magma. Morteani et al. (2013) demonstrated this finding with oxygen, sulfur, strontium and neodymium isotopes in localities in southern Madagascar; Martin (in prep.) has acquired oxygen isotopic data showing crustal values not only in calcite but also in forsterite, spinel, and phlogopite at the **Parker** phlogopite mine, Notre-Dame-du-Laus, Québec, Canada.

Many investigators, strongly influenced by the conclusions of Shaw et al. (1963), still attribute a skarn origin to both Bear Lake and Otter Lake occurrences, but the addition of new data

Fluorapatite in calcite, 9.5 cm tall
Comet Mine, Wilsons Corner, Wakefield Township, Québec, Canada
Canadian Museum of Nature collection (cat. no. 55610, ex Pinch)
Michael Bainbridge photo

increasingly supports the melting of crustal carbonates. Similar occurrences of apatite-bearing silicocarbonatite should be expected in orogenic belts marked by a major continent-continent collision. Their emplacement comes after the collision, at a stage during which there is localized metasomatism and melting of marble in response to additions "from below."

Other specimen-producing crustal silicocarbonatites may include the intrusions at **Slyudyanka**, Lake Baikal area, Russia (see page 103) and those near **Manakara**, Fianarantsoa Province, Madagascar. A skarn-type genesis (see below) is still preferred by some researchers, and that model is reported for Slyudyanka in the chapter on Russian apatite occurrences.

Apatite in Alkaline Igneous Rocks

Alkaline igneous rocks form from magmas and fluids that are so enriched in alkali elements that they precipitate sodium- and potassium-bearing minerals not usually found in more common igneous rocks such as basalt or granite. Examples of the typical alkali-rich minerals include the feldspathoids (for example, nepheline and sodalite), aegirine, and sodic amphiboles. One of these alkali elements, sodium, can substitute for calcium in apatite if other elements, such as the rare-earth elements, are present and available for coupled substitutions, and they generally are in alkaline igneous rocks. Thus a variety of apatite-supergroup minerals including fluorapatite, belovite, fluorcaphite, fluorstrophite, and britholite can be found in these rocks.

Fluorapatite is the ore-forming constituent in each of the

eleven deposits exploited in the **Khibiny** Massif on the Kola Peninsula, Russia. In contrast, the fluorapatite in the neighboring **Lovozero** Massif is widely disseminated as an accessory phase, but nowhere does it attain ore grade (Pekov 2000). Although both complexes formed at the same time (roughly 380 million years ago, Lower Devonian) and are obviously related, the reason Khibiny is so strongly enriched locally in phosphorus remains an unsolved aspect of the genesis of these huge massifs.

The ijolite-urtite alkaline igneous suite (exotic alkaline rocks with abundant nepheline) at **Khibiny** hosts the largest igneous deposits of apatite in the world (see page 103). These are a source for fertilizer phosphate that is so significant that a local town—Apatity—is named for the dominant apatite ore.

HYDROTHERMAL APATITE

In terms of morphologically fine apatite crystals, hydrothermal environments are the most significant specimen producers. This includes granitic pegmatites, described above. Hydrothermal fluids can arise from many different processes, including exsolution from a magma (magmatic) and release from hydrous minerals, such as phyllosilicates, during regional or contact metamorphism (metamorphic).

Where minerals that form from hydrothermal fluids may be disseminated throughout the country rock through which the fluids have moved. This is commonly accompanied by alteration and the formation of metasomatic rocks such as greisens, skarns and serpentinites. Superimposed on disseminated mineralization may be pathways of preferential fluid flow, such as fractures, and fracture zones.Where hydrothermal precipitation occurs along fractures, veins are formed (pictured below left), in many cases with void spaces or pockets. Mineralization may also be concentrated in pre-existing cavities or fissures (clefts). Minerals form in water-filled spaces, they have the potential to be found as well-formed crystals of striking beauty.

Magmatic Hydrothermal Systems

Hydrothermal fluids can exsolve from magmas of various compositions and under different environmental conditions, resulting in a range of possible solutes, as elements from the magma partition into the fluid phase. There is essentially no restriction on the possible rock types with which magmatic hydrothermal fluids may interact. Consequently the number of solutes that may potentially be assimilated during fluid-rock interactions is also large. Thus, the paragenesis of mineral deposits formed by precipitation from magmatic hydrothermal fluids is quite variable.

A number of different hydrothermal features are frequently associated with silica-rich granitic intrusions. In the final stages of granite crystallization from magma, the last bit of melt may become enriched in volatile elements and metals. Depending upon the degree of enrichment, the melt may transition to a fluid-dominated system. During this stage of crystallization, pegmatite formation may also occur. At some point, fluids may physically separate from the melt and form metasomatic and hydrothermal minerals in the vicinity of the intrusion. An important type of hydrothermal alteration of the parent granite is known as greisen, a metasomatic-hydrothermal rock that results from alteration by fluorine-rich fluids (Menzies 1995, Scherba 1970). Metasomatism may also extend into the surrounding country rocks. An important example of this is the formation of skarn, a metasomatic rock dominantly composed of calcium silicate minerals such as pyroxene, garnet, vesuvianite, wollastonite, and actinolite (Rakovan 2003). They are also generally enriched in iron, resulting in accumulations of magnetite and hematite. Skarns most commonly form at the contact of a silicate magma and carbonate sedimentary rocks into which it is intruded.

Skarns may host economic concentrations of ore minerals for iron, copper, lead, zinc, tungsten, tin, and additional metals. Other minerals, usually gangue, that may be present in skarn deposits include tourmaline, topaz, fluorite, corundum, beryl, and apatite. Fluorapatite, commonly with a significant chlorine content, is found in fine specimens from iron skarn deposits around the world; particularly notable occurrences are found in **Dashkesan**, Azerbaijan, among other places in the former Soviet Union (see page 103).

Skarns can also form metasomatically at the contact between carbonate- and silicate-rich rocks during metamorphism. This has long been proposed to explain calcite-hosted apatite in the **Grenville Province** of southwestern

Left:
Hydrothermal vein showing black wolframite crystals in quartz
Yaogangxian Mine, Hunan Province, China
Berthold Ottens photo

Facing page:
Apatite, 7.7 cm tall
Snarum, Buskerud, Norway
Houston Museum of Natural Science collection
Jeff Scovil photo

Québec and Ontario and at **Slyudyanka**, Baikal area, Russia. A more recent model for their formation, crustal silicocarbonatites, is described above.

Vein and greisen tin-tungsten deposits are some of the most important examples of apatite-bearing igneous hydrothermal systems. These exhibit a close association with granitic plutons, especially those that have formed from late-stage, highly evolved melts. Famous examples include those associated with tin porphyry deposits of the Bolivian Tin Belt (**Llallagua** and **Oruro**; see page 67); **Cornwall-Devon**, southwest England; **Panasqueira**, Portugal (page 70); **Pasto Bueno**, Peru; **Aktschatau**, Kazakhstan (page 103); and granites of the Erzgebirge in Germany and the Czech Republic: **Zinnwald**, **Altenberg**, **Horni Slavkov (Schlaggenwald)**, **Ehrenfriedersdorf** (page 18). Also, most of the tungsten deposits of southeastern China, the richest tungsten province in the world, occur in veined and greisenized granite and associated skarns (**Yaoganshan**, **Piaotang**, and **Wutong** mines, among others; page 98).

In some cases, clay minerals such as kaolinite, rather than larger mica crystals, form during greisenization. In such cases, the metasomatic alteration is refered to as *kaolinitization*. These deposits can be major sources of high-quality kaolin such as the china-clay deposits of **Cornwall-Devon**, United Kingdom, that are directly related to the tin greisens in the same area. The apatite-bearing kaolinite deposit at **Fubasami** Tochigi-ken, Japan (page 98), formed in the same way.

Vein and metasomatic deposits dominated by other metals, including copper, silver, gold, lead, and zinc, may be associated with hydrothermal fluids that have evolved from magmas of various compositions. Examples of such deposits, in which apatite specimens are significant, include the **Ashio Mine**, Japan (copper); the **Julcani** and **Huanzala** mines, Peru (lead-zinc); the **Sweet Home Mine**, Alma, Colorado and the **Wutong Mine**, Liubao, China (both lead-zinc-silver).

Metamorphic Hydrothermal Systems

In mountain-building events like the Alpine orogeny, rocks are structurally deformed and metamorphosed owing to the stress and subsequent increases in pressure and temperature. Structural deformation of brittle rocks leads to breccias, fractures, faults, and openings within them. Fissures created by local tensional or shear (transpressional) stresses are also known as "clefts." Metamorphic hydrothermal fluids are created by the dehydration of hydrous minerals and released during the compression of sedimentary rocks. They pick up solutes as they circulate through and react with regional metamorphic and igneous rocks. Where they encounter fissure systems, precipitation will occur in response to decreasing pressure and temperature. These mineral-filled cavities are known as *Alpine clefts*.

Johann Georg Koenigsberger (1874–1946) was the first to note that the paragenesis of Alpine cleft minerals reflects the composition of the proximal host rocks and that the solutes required for mineral formation originate near the clefts. Thus, clefts hosted by amphibolites and other metamorphic lithologies exhibit different mineral assemblages than those encountered in igneous intrusions. Silica-rich felsic intrusions and their metamorphic equivalents commonly host quartz-dominated clefts and may contain apatite. Those found in

mafic and ultramafic igneous and metamorphic rocks commonly host such minerals as vesuvianite, diopside, apatite, titanite, iron-titanium garnet, and chlorite. Characteristic cleft minerals in mica-rich metasediments such as schists and gneisses include anatase, rutile, brookite, hematite, magnetite, and ilmenite (Graeser 1998).

Apatite crystals from Alpine clefts are usually tabular in habit. They are often quite complex, exhibiting many different forms, so much so that they may develop a rounded habit. They are generally colorless or only very lightly tinted and are non-fluorescent. A common aggregate habit consists of multiple tabular crystals that are intergrown in a parallel fashion but slightly offset from one another, creating a stair step-like effect.

As other common Alpine cleft species including quartz and K-feldspar (adularia variety), apatite may be found as faden crystals because mineralization and cleft formation generally occur simultaneously over extended periods of time (Richards 1990). Fibrous amphibole, often referred to as "byssolite," is commonly reported as an inclusion in apatite from Alpine clefts hosted in amphibolite and other metasedimentary rocks.

The classic occurrences of Alpine clefts are in the Swiss and Austrian **Alps**; other major examples are found in the **Himalayas** of Pakistan (see page 86) and Afghanistan (page 96). Two extraordinary Alpine cleft deposits are the **Dodo** and **Puiva** mines in the sub-Polar Urals, Russia (page 103). Several important occurrences also exist in the Appalachian fold-and-thrust belt of North America, including the **Keystone Quarry**, Cornog, Pennsylvania, and **Acushnet**, Massachusetts.

Apatites of the United States

Carl A. Francis and Donald A. Dallaire

Apatite is the most common phosphate in the earth's crust, occurring in diverse geological environments (see page 21). Of the three members of the apatite subgroup, only fluorapatite is common. Chlorapatite is unknown in hand specimens from the United States, but hydroxylapatite is probably more common than reported because apatite is easily recognized and is not routinely analyzed.

GRANITIC PEGMATITES

Fluorapatite is common in many granitic pegmatites and may appear at least three times in the crystallization sequence. It first crystallizes as a magmatic mineral usually carrying some manganese in substitution for calcium (variety manganapatite), which makes it fluoresce yellow under shortwave ultraviolet radiation. Crystals and masses are locked in matrix, usually feldspar. Toward the end of magmatic crystallization, a hydrothermal solution may evolve that dissolves solid pegmatite and deposits minerals in the newly created open space. Fluorapatite is a common mineral in pockets lined with albite in the platy habit known as "cleavelandite." Here apatite is found as freestanding, lustrous, translucent-to-transparent hexagonal crystals, which range in proportion from tabular to prismatic. They may be colorless or colored—usually green, blue, or purple. These beautiful crystals rarely reach dimensions of 5 centimeters. Still later in the paragenetic sequence, apatite may crystallize as microcrystals or a fine-grained coating on feldspar and other minerals lining pockets. This final generation of apatite may contain carbonate and be low in fluorine. Of the two distinct families of granitic pegmatites, NYF and LCT, apatite specimens are rarely found in NYF pegmatites.

Maine

The many apatite occurrences in Maine's pegmatites and other rock types are described in detail by King and Foord (1994); only the most prominent can be mentioned here. Pegmatites in the towns of Auburn (Androscoggin County), Hebron, Paris and Greenwood (all Oxford County), which make a counterclockwise arc around the northeast margin of the Sebago batholith, are famous for crystals of fluorapatite as well as tourmaline.

The **Pulsifer Quarry** in Auburn is the premier locality for fluorapatite in the United States. The apatite's intense "royal purple" color has made crystals from the Pulsifer Quarry world classics. In 1901 Pitt P. Pulsifer discovered a pocket that contained purple crystals thought to be tourmaline. These were sent to Harvard University, found to be fluorapatite, and described by Wolff and Palache (1902). Pulsifer continued working his small quarry on a limited basis and years later mined the best purple apatite he ever found. Sometime in 1918 this approximately 4-by-4-by-3-centimeter specimen was purchased by Mrs. Roebling as a birthday present for her husband Colonel Washington A. Roebling for $250! A decade later the Smithsonian Institution acquired the specimen as part of the Roebling collection; it remains part of the National Mineral Collection. The "Roebling Apatite" is the most famous apatite specimen in the world.

Right: Donated to the University of New Mexico by Arthur Montgomery, the Harding Quarry near Dixon, Taos County, New Mexico yields fluorescent, blue fluorapatite crystals associated with spodumene, lepidolite, and quartz.

Facing page: True to its name "the deceiver," this unusual 4-cm-tall fluorapatite on albite ("cleavelandite") looks a lot like elbaite, but a simple hardness test easily differentiates these two very different mineral species. Himalaya Mine, Mesa Grande San Diego County, California Bill Larson collection Robert Weldon photo

Upper left: Washington A. Roebling (1837–1926) in 1854

Upper right: The "Roebling Apatite," 4 cm wide
Pulsifer Quarry, Mount Apatite, Auburn, Maine
Smithsonian Institution collection, Chip Clark photo

Below: Fluorapatite and quartz, 1.3 cm wide crystal
Mount Rubellite, Hebron, Maine
Jim Mann collection, Jeff Scovil photo

Much of the top of Mount Apatite is underlain by pegmatite. The Pulsifer pegmatite is nearly flat lying but has been worked over the years as separate quarries. Following Pulsifer the quarry was not worked much until 1964 and 1965 when a lease was obtained by Irving "Dudy" Groves. In 1966 Terry Szenics took over the lease for two years and with the help of Frank C. Perham opened pockets that greatly enhanced the quarry's reputation (Szenics 1967). The adjacent **Dionne**, **Groves**, and **Wade** quarries exploit the same pegmatite and have also yielded fine purple apatite specimens.

Mount Rubellite in Hebron is a minor tourmaline locality that produced locally famous fluorapatite with a distinctive habit, comprised of a low dipyramid and practically no prism or pinacoid faces. Jim Mann made a major find of colorless, green, and purple tabular crystals at the locality in 1985.

Mount Mica, the famous gem tourmaline locality in Paris, rarely produced apatite specimens. However, a drawing of a small, deep blue, marvelously complex crystal with new and unusual crystal faces is reproduced in Dana (1892). Then in 2003 Gary and Mary Freeman opened a pocket of extraordinary, large deep blue crystals. More recently they encountered another pocket area, which yielded both prismatic and tabular apatite crystals up to several centimeters. The color of these crystals ranges from colorless to light blue, to pale pink, with some crystals exhibiting pronounced color zones. Unfortunately many were coated with an unidentified mineral, which in many cases cannot be removed.

Through the years, the **Bennett** and **Orchard** mines in Buckfield have also sporadically produced specimens of blue to green apatite that generally reside in local collections. Dennis Durgin's recent efforts at **Mount Marie**, also in the town of Paris, have been rewarded with fine gem tourmaline and important discoveries of purple apatite. With these finds, Mount Marie rises from a previously obscure apatite-producing locality to one of significance.

The **Harvard Quarry** in Greenwood, Maine, is the state's second-best known purple apatite locality. Over the decades, the Harvard Quarry has consistently produced purple (albeit a

Upper left: Fluorapatite
1.6 cm long crystal
Mt. Mica, Paris, Oxford County, Maine
Larry Stifler collection, Jeff Scovil photo

Upper right: Fluorapatite, 4 cm wide
Durgin Prospect #1, Mt. Marie
Paris, Oxford County, Maine
Dennis Durgin collection, Jeff Scovil photo

Right: Fluorapatite on quartz with elbaite
11.2 cm wide
Harvard Quarry, Greenwood
Oxford County, Maine
Don Dallaire collection
Kevin Downey photo

Lower right: Fluorapatite, 3 cm wide
Emmons Mine, Greenwood
Oxford County, Maine
Emmons Mine specimen
Kevin Downey photo

paler shade than those from the Pulsifer), blue, and green crystals. Collectors still can find attractive specimens sporting crystals up to 6 millimeters on the old dumps.

On the other side of the valley, Raymond Sprague works the **Emmons Mine** for mineral specimens. Colorless, blue, and purple fluorapatite crystals to one-centimeter-long are routine for Sprague, but a distinctive find of zoned, teal-blue crystals reaching 3 centimeters was found in 2001.

Beryl-bearing non-lithium pegmatites in the **Albany** area have yielded fine quality colorless, green, and blue fluorapatite specimens. Further north the important pegmatites at **Newry** have yielded fluorapatite in various habits and colors. Most notable are dense druses of pale purple prisms lining veins cutting through quartz from the **Bell Pit**. A significant find of stout crystals to 3 centimeters in length was recently made by Louise Jonaitis and Bob Brown in the **Crooker Pit Extension**. They are translucent, green-gray in body color and coated with a blue-white film.

New Hampshire

Notable localities for magmatic "manganapatite" in New Hampshire are the **Ruggles Mine** in Grafton, Grafton County, where it is sometimes graphically intergrown with microcline and the **Parker Mountain Mine** in Strafford, Strafford County, where it is vividly fluorescent. The **Keyes Mines** on Mount Tug in Orange, Grafton County, produced New Hampshire's finest apatite specimen. Crystals from this mine typically occur in vuggy albite with muscovite and schorl. They vary in color from blue to green to light purple and may be zoned. Habits range from dipyramidal to prismatic and sizes reach 2 centimeters. A former collector and mineral dealer, Bill Brownell obtained an exceptional specimen in 1961 while visiting the then-operating mine. The mica miners had opened a fine pocket, and he purchased a 20-by-20-centimeter specimen exhibiting sharp blue apatite crystals up to 2.4 centimeters wide with muscovite crystals on albite. Brownell mentioned at the time that the pocket was never fully excavated, and it is likely other specimens were left behind.

The **Palermo # 1 Mine** located in Groton, Grafton County, is a world-famous phosphate pegmatite that has yielded at least 140 species, and it is the type locality for 10 of these (Whitmore and Lawrence, 2004). The Palermo #1 Mine is New Hampshire's most diverse apatite locality. Fluorapatite is found in all of the pegmatite's zones, but the best specimens occur in the wall zone and in hydrothermal veins along the margin of the quartz core. While most of the crystals are 1 to 3 millimeters in size, many examples are in the one-centimeter range, with the occasional crystal reaching 10 centimeters. Colors range from pure white to light pink to green to blue. In the late 1980s Robert Whitmore and his crew were working underground when they encountered an apatite mass that was approximately 2 meters in diameter. Abundant white, almost

Left: Fluorapatite, 3.5 cm tall
Stickland Quarry, Portland, Middlesex County, Connecticut
Edna Whitmore collection, Kevin Downey photo

Right: Fluorapatite "stack," 3.7 cm tall
Palermo #1 Mine, North Groton, Grafton County, New Hampshire
Don Dallaire collection and photo

Facing page: Fluorapatite on muscovite and albite, 21.3 cm tall
Specimen acquired by Bill Brownell in 1961
Keyes Mine, Grafton County, New Hampshire
Don Dallaire collection, Jeff Scovil photo

porcelaneous crystals occurred in an extraordinary number of habits in pockets in the massive apatite, the surrounding albite, and in nearby quartz pockets. Individual crystals range from 1 to 15 millimeters but often formed as tabular individuals in distinctive stacked "chains" up to 10 centimeters long. Crystals also formed as thin hollow tubes resembling spaghetti and as plumose aggregates. Apatite crystals are also commonly found with other phosphate minerals.

Connecticut

Both magmatic and hydrothermal fluorapatite is common in the Middletown district of central Connecticut. At the now-reclaimed **Gotta-Walden** and **Strickland** quarries, both in the town of Portland, colorful magmatic masses of green manganese-rich apatite were common and were often found in association with deep red almandine masses. A small number of specimens with purple crystals to 3 centimeters were found. Their color matches that of the best Pulsifer Quarry specimens, making Strickland Connecticut's premier apatite locality.

The **Gillette Quarry** at Haddam Neck is famous for freestanding elbaite, but colorless to pale green, pink, or blue tabular apatite crystals to 2 centimeters in diameter occurred in the pockets as well. The **Fillow Quarry** at Branchville in southwestern Connecticut is famous as the type locality for lithiophilite and seven other manganese phosphates. Fluorapatite occurs there as well but is overshadowed by the type minerals. A wonderfully complex drawing of a pocket apatite from Branchville appears in Dana (1892).

North Carolina

Granitic pegmatites occur throughout the Appalachian Mountain chain as far south as Alabama, but few outside of New England are phosphate-rich. A notable exception is the **Chemetall Foote Corporation's Lithium Mine** (aka **Foote Mine**) near Kings Mountain, Cleveland County, North Carolina—the type locality for ten species! Fluorapatite crystallized with other late-stage species on fractures in massive pegmatite. Crystals are colorless, beige, blue, or purple and vary in habit. They are frequently small and make fine micromounts, but crystals may reach one centimeter, and cabinet specimens of druses are not uncommon.

South Dakota

The Black Hills in southwestern South Dakota have a rich mineralogy (Roberts & Rapp 1965) and are especially famous for their phosphate-rich pegmatites. The most prolific apatite locality in the Black Hills, the **Hugo Mine** near Keystone, Pennington County, is remarkable for its giant crystals as well as its diversity of hydrothermal occurrences. Roberts & Rapp (1965) report that a crystal, showing sharp prism faces in cross section, was estimated to weigh one ton! They also describe seven distinct apatite occurrences at the mine that vary in crystal or aggregate habit, color, and association. Notable primary fluorapatite also occurs at the **Bull Moose**, **Hot Shot**, and **November** mines. The **White Cap Mine** at Keystone yielded gemmy, amethyst-purple crystals, the largest of which,

> *... measures ⅞ x ⅞ x ⅝ inch in size, has short prism faces, large basal pinacoid and first order pyramid faces, and rivals the finest crystals from any locality for color and crystal perfection.* (Roberts & Rapp 1965)

But the Black Hills apatite locality probably best known to

collectors is the nearby **King Lithia Mine**. Handsome hemispherical clusters of purple prisms on feldspar are King Lithia classics. Roberts & Rapp (1965) mention stubby doubly terminated olive green crystals on milky quartz and colorless rosettes to 2 centimeters. Marc and Debra Wilson made a distinctive find of blue and white-zoned crystals in 1981 (Wilson et al. 1984).

Idaho, Colorado, New Mexico, and California

Two lustrous, 5-centimeter fluorapatite crystals labeled "Idaho" and "Boise Co., Idaho" in the Harvard University collection were supplied by the famous field collector Arthur Montgomery in the 1930s. They are probably from the miarolitic **Sawtooth** batholith and suggest potential for future finds in this area.

Fluorapatite is known from Colorado's rare element LCT pegmatites but not from its famous amazonite-bearing NYF pegmatites west of Colorado Springs. The **Crystal Lode** pegmatite near Vail in Eagle County is Colorado's best fluorapatite locality. Specimens from this locality are usually isolated, 1- to 3-centimeter long, translucent prisms, which show a yellow color that is atypical for pegmatitic apatite. The **Harding** pegmatite near Dixon, Taos County, New Mexico, is owned by the University of New Mexico and is consequently one of the world's best-studied pegmatites. It hosts fluorescent, blue fluorapatite crystals of 5 centimeters or more, embedded in an association of spodumene, lepidolite, and quartz.

Fluorapatite is only a minor associate of tourmaline in the famous pegmatites of San Diego County, California. Thumbnail and miniature-size specimens are uncommon but available. Colorless, blue, and especially pink prismatic crystals up to several centimeters come from the **Himalaya Mine** at Mesa Grande. The mines at **Pala**, the **Little Three Mine** at Ramona, the **Maple Lode Mine** on Aguanga Mountain, and the **Pack Rat Mine** at Jacumba have all supplied fine specimens. Particularly desirable are apatite crystals perched on elbaite or cleavelandite.

ALPINE CLEFTS AND HYDROTHERMAL VEINS

Alpine cleft mineralization, which is restricted to metamorphic terranes, is uncommon in the United States. Perhaps the best occurrence is the crushed stone quarry in **Acushnet**, Bristol County, Massachusetts, which is currently owned and operated by the P.J. Keating Company. Clefts are lined with albite crystals of the pericline habit that may reach 10 centimeters in length and are typically coated with dark green flaky chlorite. Notable associates are adularia, calcite, epidote, fluorapatite, and titanite, but only the fluorapatite rises to world-class quality. Equant crystals are commonly translucent gray to one centimeter nestled among chlorite-coated albite crystals. Exceptional crystals exceed 10 centimeters; yellow prismatic crystals also occur.

The **Keystone Quarry** at Cornog, Chester County, Pennsylvania, is another cleft locality, this one hosted in amphibole and biotite gneisses. The Biggs family found a few fluorapatite specimens in the quarry in the late 1960s. One is a 2.3-by-1.8-by-1.4-centimeter compound crystal with sparse inclusions of actinolite that is perched on a blade of clinozoisite. It displays simple prism and pinacoid faces that are modified by narrow dipyramidal faces. Two larger specimens are interesting for their overall aggregate habit. Both are groups of crystals in parallel position that are offset and resemble stair steps; a common habit for apatite from Alpine clefts elsewhere in the world. These are 5 to 7 centimeters across, pale in color, transparent, and associated with fibrous actinolite (variety "byssolite").

Above: John McLean working a pegmatite at the Himalaya Mine, San Diego County, California in the 1980s
Photo courtesy of Pala International

Below left: Fluorapatite, 3.6 cm tall
Crystal Lode, Fulford, Eagle County, Colorado
Dave Bunk collection, Jeff Scovil photo

Below middle: Fluorapatite, 5.1 cm tall
Packrat Mine, San Diego County, California
Cal and Kerith Graeber collection, Jeff Scovil photo

Below right: Fluorapatite, 2.7 cm wide
Keystone Quarry, Cornog, Chester County, Pennsylvania
C. Carter Rich specimen, Jeff Scovil photo

Facing page: Fluorapatite, 11.5 cm tall
Acushnet Quarry, Bristol County, Massachusetts
Mike Haritos collection, Jeff Scovil photo

Hydrothermal vein occurrences such as those at the **Jeffrey Quarry**, in Pulaski County near North Little Rock, Arkansas, yield fine crystals but few matrix specimens. Because analyses are not available for most occurrences, some may prove to be hydroxylapatite. A road cut in **Marlboro**, Middlesex County, Massachusetts, produced 1- to 2-centimeter-long yellow prisms. Drawings in Dana (1892) depict complex crystals from **Hiddenite**, Alexander County, North Carolina. Fine crystals up to one centimeter and larger crystals that are commonly obscured by chalcopyrite come from the **Eagle Mine**, Gilman, Eagle County, Colorado. The **Sweet Home Mine** at Alma, Park County, Colorado, yielded quartz, rhodochrosite, and sulfide matrix specimens scattered with lustrous prismatic apatite crystals to about 5 millimeters.

MARBLES AND SKARNS

Fluorapatite occurs in the Grenville age marbles, skarns, and vein dykes of St. Lawrence County, New York; however none of the localities is as prolific or produces crystals as large as the world-class occurrences in adjacent Ontario and Québec (see pages 38). The premier New York state locality is the **Rose Road** wollastonite occurrence (formerly known as the **Mulvaney Prospect**) at Pitcairn, which yields bright blue crystals to 15 centimeters. Apatite is also found at **Bush Farm**, the classic brown uvite locality, in Gouverneur. Green to blue-green crystals occur near the **Macomb** "peristerite" (a sodic plagioclase) locality. In the past decade Mike Walter recovered abundant pale green-gray crystals to 10 centimeters at **Powers Farm**, the classic, black uvite locality in Pierrepont, New York. Attractive blue crystals are found at **Clarks Hill** and elsewhere in **Rossie**, and large gray-blue crystals come from **Moon Lake** in Theresa, in neighboring Jefferson County.

Fluorapatite crystals similar to those found in St. Lawrence County occur in the orebodies and enveloping Franklin Marble at **Franklin** and to a lesser extent **Ogdensburg**, both in Sussex County, New Jersey (Dunn 1995). Notably arsenic-rich, crystals and masses from the orebodies are typically blue, blue-green, or green and are variably fluorescent.

IRON DEPOSITS

Fluorapatite is frequently associated with magnetite in iron deposits, where it is an economically undesirable gangue mineral because phosphorus can make the resulting iron or steel brittle. Only a few of the significant occurrences of fluorapatite from iron deposits are of interest to collectors. For example, anhedral grains of fluorapatite are abundant in the granular and massive magnetite in the **Mineville** iron deposit in Moriah Township, Essex County, New York, a Kiruna-type deposit of magmatic origin. They are scientifically interesting because they contain high concentrations of thorium and rare earth elements substituting for calcium and are associated with a suite of rare earth minerals but they are not of display quality (Luo et al. 2011).

The **Pea Ridge Mine** at Sullivan, Washington County, in southeastern Missouri, is another magmatic Kiruna-type iron deposit. Fine-grained magnetite and subordinate specular hematite orebodies intrude Precambrian age porphyries (Emery 1968). The deposit is so rich in fluorapatite that the mineral has been recovered as a byproduct for phosphorus. The magnetite zone is quite vuggy, and the cavities are lined with quartz, specular hematite, and apatite. The brown crystals are prismatic with pyramidal and occasionally pinacoidal terminal faces. They are reminiscent of crystals from the marbles and skarns in New Jersey and New York, although those from the Pea Ridge Mine only reach a few centimeters in length. Pea Ridge is the only significant apatite specimen locality in the mid-continent region. Mining activity has recently started again at the locality, but now for rare earth elements, which are enriched in the apatite found there (Long et al. 2010).

Magnetite and hematite replaced limestone surrounding laccoliths of quartz-monzonite porphyry in the **Iron Springs** district near Cedar City in Iron County, Utah (Mackin 1968). These deposits produce single crystals and matrix specimens of yellow fluorapatite that are similar to the familiar specimens from Cerro de Mercado in Durango, Mexico (see page 48).

Magnetite octahedrons altered to hematite occur in seams in rhyolite at the famous "martite" locality at **Twin Peaks**, southern Millard County, Utah, and small white to green-yellow prisms of apatite and black prisms of augite are associated with them. This locality seems analogous to the contemporary magnetite locality at Huanaquino Mountain near Potosí, Bolivia (see page 67).

HYDROXYLAPATITE

Holly Springs in Cherokee County, Georgia, is the best described hydroxylapatite locality in the United States (Mitchell et al. 1943). The locality is a "verde antique" (i.e., serpentine) quarry. Yellow transparent apatite occurs as rounded masses and prisms to 2 centimeters in length embedded in pale green talc and chlorite schist. Specimens are scarce and only suited for systematic collections and research use (see page 7).

Yellow apatite also occurs in a talc prospect near **Nemo** in the Black Hills of South Dakota (Roberts and Rapp 1965). At both the Holly Springs and Nemo deposits, the talc is certainly a contact metasomatic rock formed between serpentine or other ultramafic rock and a more silicic country rock, perhaps schist. An analogous biotite rock at the anthophyllite asbestos mine in **Pelham**, Hampshire County, Massachusetts, contains isolated masses and crystals of corundum and green hydroxylapatite. Undoubtedly similar mineralization occurs elsewhere and should be investigated.

CONCLUSION

The outstanding localities for fluorapatite specimens in the United States are primarily granitic pegmatites, most notably those in southwestern Maine. The Pulsifer pegmatite in Auburn is exceptional for its intensely purple crystals. Pegmatites in other parts of the country also yield fine specimens. Occasionally specimens come from Alpine clefts, hydrothermal veins, and marbles across the U.S. Quality hand-sized specimens of hydroxylapatite are surprisingly unknown in the vast and geologically varied United States of America.

ACKNOWLEDGEMENTS: *The authors wish to thank the following people for sharing their knowledge of U.S. apatite occurrences: Miranda Boenisch, Steven Chamberlain, Robert Cook, Jesse Fisher, Carlton Holt, Anthony Kampf, Thomas Loomis, James Mann, Alyssa Morgan, Harris Precht, Gary Smith, Robert Whitmore, Marc Wilson, William Wray, and everyone who posted photographs on mindat.org.*

Top left: Fluorapatite and quartz, 2.6 cm tall
Jeffrey Quarry, North Little Rock, Arkansas
Jimmy McNeil collection, Jeff Scovil photo

Above: Fluorapatite on tremolite, 9 cm tall
Bush Farm, St. Lawrence County, New York
Don Dallaire collection, Kevin Downey photo

Right: Fluorapatite and quartz, 2 cm tall
Sweet Home Mine, Alma, Colorado
Dave Bunk collection, Jeff Scovil photo

Lower right: Fluorapatite on magnetite
crystals to 1.3 cm long
Cedar City, Iron County, Utah
Seibel Minerals specimen, Jeff Scovil photo

Lower left: Fluorapatite, to 1.5 cm wide
Eagle Mine, Gilman, Eagle County, Colorado
Bill Warren collection, Jeff Scovil photo

Left: Fluorapatite, 6.8 cm tall
Pea Ridge Mine, Washington County, Missouri
Laurence M. Nuelle collection, Jeff Scovil photo

Thousand-Pound Crystals and Eighty-Carat Gemstones–Canada is BIG!

Kimberly Tait and Michael Bainbridge

In 1821 James Wilson, a young medical doctor from Scotland, came to Canada to set up practice in the small town of Perth in southeastern Ontario. He was an enthusiastic student of geology and quickly became an expert on the rocks and minerals of his new surroundings. Dr. Wilson is probably best known for his discoveries of "wilsonite" and "perthite." In this context, however, his most notable contribution was his observance of large quantities of "phosphate" in the metamorphic rocks of the Perth area.

Perth lies at the edge of the Central Metasedimentary Belt (CMB) of the Grenville geological province, one of the most geologically varied and interesting regions in Canada. The CMB is rich in mineral deposits and exceptional crystals of, among other things, apatite.

But in 1821 the CMB was the short straw if you were a farmer coming to Canada on the promise of land. Unlike the arable limestone plains to the south and east, hardly anything would grow in the thin, rocky soils of the CMB—ironic, considering that apatite is the primary source of the phosphate used in fertilizer.

Intent on mining phosphates, Dr. Wilson obtained the rights to several properties along the eighth concession in Burgess Township in 1852. Canada's first apatite mine was recorded when work began on Lot 2 in 1855. Shortly thereafter, apatite "mines" started cropping up in farmers' fields. Few of those mines ever amounted to anything more than a hole in the ground in somebody's back forty, but a phosphate rush had begun that would soon sweep throughout the CMB from southeast Ontario to southwest Québec. A large processing plant was built in Buckingham, Québec, and Canada quickly became a major international supplier.

By the turn of the century, however, cheaper sources of phosphate were found in large sedimentary deposits in Florida, so Canadian production slowed considerably. Mica had always

Above: Fluorapatite
Specimen 18 cm long, cut stone 24.26 carats
Liscombe Mine, Haliburton County, Ontario
Canadian Museum of Nature collection (cat. nos. 31187, 20828)
Michael Bainbridge photo

Facing Page: Map of Canada, William W. Besse cartography

been an important secondary commodity in these deposits though. At first only large, transparent sheets for use in windows were recovered, but with the rise of the electrical industry and mica's application as an insulator, it soon far surpassed apatite as the primary commodity of the phosphate mines. As demand rose, the General Electric Company acquired direct stakes in a number of the mines. What was once a backyard industry that helped to supplement local farmers' agricultural income was now a growing concern.

But by the 1950s, commercial production of both apatite and mica had effectively ceased as other, less expensive sources for mica had become available. Still, the innumerable pits and trenches left in the wake of the industry set the stage for what was to come, as the outdoor hobby of mineral collecting—rockhounding—was growing in popularity.

In 1967 the Ontario Department of Mines reported *total* apatite production for the entire province at $394,519 (1870~1951), but by 1995 visiting rockhounds were bringing more than $5,000,000 (Canadian) into the Bancroft economy *annually* (Fouts 1995).

SPECIMEN PRODUCTION

Apatite from the Grenville province has been sought after by collectors since at least when Dr. Wilson first began poking around roadside outcrops in between house calls. In the *Report of the Royal Commission on the Mineral Resources of Ontario of 1890*, W.J. Morris recounts:

> *The finest crystals in the world were got* [north of Kingston]*; I got one that weighed 1,000 lb. I have seen hundreds of crystals weighing from 50 to 100 lb. In loading cars we have had to break every crystal with the hammer for fear they would be stolen as specimens.*

Few specimens are known to have survived these early days of mining, but good examples can be seen in the collections of area institutions such as the local museum in Perth, where Dr. Wilson's collection resides.

Apatite has now been found in a wide range of geological environments all across the country, but without a doubt, the Grenville province remains the country's most prolific and best known apatite specimen producer. Contributing to this prolificacy is the fact that the area is well enough developed and close enough to major population centers to allow easy access, but much wilderness remains where many of the historic mica/apatite mines can still be found. Certain localities have been popularized in the international marketplace thanks to the work of a handful of industrious dealers, but the vast majority of specimen production is accounted for by field-

collectors who extract crystals by hand from dozens of scattered localities few have ever heard of.

For the individual collector, there is hardly a bad place to dig, but there are a few standouts. When speaking of the great worldwide localities for apatite specimens, the Yates and Liscombe mines must undoubtedly be mentioned, but there are also exceptional lesser known localities in other parts of the country. Canada's extensive northern region, for example, has shown great potential.

CENTRAL METASEDIMENTARY BELT

The full exposed extent of the Grenville geological province in Canada spans nearly two thousand kilometers from southern Ontario to the east coast of Labrador, so it should be noted that (as with the phosphate mining industry) the area of interest to the collector is almost exclusively limited to the Central Metasedimentary Belt. Granted, "Grenville" has a nicer ring to it, but the distinction between the belt and the entire province is made for two reasons: the CMB is exposed only at the southern portion of this great expanse, which narrows the focus to a small part of southeastern Ontario and southwestern Québec (see map page 39); and Grenville is best known among collectors for the calcite vein-dikes, which are predominantly found in the CMB.

These calcite-cored pegmatites, skarns, marbles, carbonatites, crustal silicocarbonatites, or whatever they are called (each may be accurate in at least some instances) are a rockhound's dream. Twelve-thousand years of erosion since the last glaciation, and the thin veneer of topsoil that was once the bane of the early settlers' existence now means easy access to loose crystals in earth-filled cavities.

Apatite from the CMB is most commonly various shades of green or red (sometimes both in the same crystal), but it can also be brown, black, blue, nearly white, and occasionally colorless. Depending on the occurrence, it can be completely opaque to totally transparent but is usually highly fractured. Good crystals in the 5- to 10-centimeter range are common at most localities, with 15- to 20-centimeter crystals found at some. And a few localities are known for producing much larger crystals.

They occur as stubby to elongated hexagonal prisms. They are commonly doubly terminated, with a somewhat "melted" appearance and are found in massive white, pink, or orange calcite. Terminations are usually rounded or "bubbly" (illustration left), but sharp crystals, often with a pronounced basal pinacoid, are found at some sites. Crystals often display parallel growth interrupted by other crystals or the enclosing calcite matrix. The apatite is frequently associated with mica, feldspar, pyroxene, and/or dozens of complex amphibole group minerals (aka "hornblende"). Titanite, scapolite, fluorite, and zircon are also common accessory species.

With only one exception (see below), all of the apatites from producing localities in the CMB are fluorapatite. And unlike the surrounding calcite matrix, they are quite resistant to acid erosion. As a result, many lustrous single crystals, floater groups (with or without associated minerals), and chunks of wall-rock (with attached crystals) are recovered from the soil in the weathered portion of a vein before the solid calcite core is reached. Despite the fact that CMB apatite specimens typically appear on the international market as stable crystals set in relief against a pristine calcite matrix, attempts to extract such specimens in the field by the casual collector often end in disappointment because of the brittle nature of the "frozen" crystals and the tenacity of the enclosing calcite (savvy collectors will increase their chances by using superglue to stabilize the crystals against breakage first). Mechanical removal of bulk vein material is usually required to obtain any quantity of good matrix specimens.

Ontario

The **Liscombe Mine**, Monmouth Township (Municipality of Highlands East), Haliburton County, was worked for mica in the early 1900s and has been locally known for quite some time. It rose to prominence in 1975, when it was worked commercially for gemstones.

Liscombe Mine fluorapatite crystals are found in light orange-pink, fine-to-coarse-grained calcite with brown phlogopite. They are generally crude, being a little more rounded than most others from the CMB, but they are well-renowned for their great clarity, comparatively low degree of fracturing, and bright green to yellow-green color. There are numerous other localities in the vicinity that have produced fine green gems weighing up to several carats, but Liscombe is the only one where the quality and consistency of the apatite is sufficient to have invited commercial production.

Cut stones were marketed under the name "Trilliumite," after Ontario's official provincial flower, the white trillium (*Trillium grandiflorum*). In *Gemstones of North America Vol. III*, the late John Sinkankas tells of a "museum special" faceted stone weighing 86.89 carats that was offered by Trilliumite Explorations Ltd. as "the largest gem apatite in the world … fine chrome green color, flawless" for US$8,700. Three other exceptionally large stones (and many smaller ones) were cut from rough netted by this venture.

Canadian dealers Rod and Helen Tyson, who leased the

Right: Fluorapatite ("trilliumite"), 36.92 carats
Liscombe Mine, Haliburton County, Ontario
Robert Beckett collection
Michael Bainbridge photo

Below: White trillium *Trillium grandiflorum*, the official provincial flower of Ontario and namesake of the trade name for faceted apatite from the Liscombe Mine
Don Komarechka photo

Facing page:
Fluorapatite, 4.3 cm tall
Canoe Lake Roadcut
Brad Wilson collection
Michael Bainbridge photo

property in the mid-90s, brought out most of the specimens recently on the market. They used a small pneumatic chisel and air abrasive to prepare the apatite. The Tysons "soak the specimens and matrix with a stabilizer then remove the post-preparation residue chemically" (tysonsfineminerals.com).

Many institutions around the world have large apatite specimens labeled as being from "**Tory Hill**" or something to that effect. Most of these will likely have come from a series of related occurrences along the Madill/Gibsons Road, the most famous being the **Bear Lake Diggings** (aka **Gibsons Road West Occurrence**), Monmouth Township (Municipality of Highlands East), Haliburton County.

Bear Lake is well-known not only for large crystals of apatite but also for exceptional examples of titanite, biotite, microcline, and "hornblende," all of which can be found in coarse white calcite. The Bear Lake Diggings also popularized the "Bancroft Collecting Area" experience by becoming the first location in the province to be protected as a public mineral collecting site.

Minerals occur in a series of roughly east-west trending vein-dikes cutting gneiss. They vary in width from a few centimeters to a few meters. Some of the veins have been excavated by collectors to depths up to 5 meters. Bear Lake apatite is usually found as stubby, translucent to transparent light green crystals with well-developed prism faces and crude terminations.

No significant commercial production is known to have taken place here, but traveling rockhounds and casual collectors continue to recover tons of good specimens every year. From time to time, the Bancroft & District Chamber of Commerce, who manages the Bear Lake Diggings as a fee-to-collect site, exposes productive new veins for collectors.

Well-formed, large, doubly terminated crystals of red/brown and green apatite are found in weathered veins at a number of sites around Lake Clear (Turner's Island, Park Mine, **Smart Mine**, **Meany Mine**, Miller Property), Sebastopol Township (Township of Bonnechere Valley), Renfrew County. The Royal Ontario Museum houses several exceptional, large crystals of bright red apatite from the **Park Mine**. These localities are also known for exceptional examples of other minerals commonly associated with apatite, including large (to about 5 centimeters) hyacinth-red zircon crystals. Historically, **Turner's Island** is perhaps the most notable locality in the area for apatite, but the one that is of the most interest to collectors today is probably the **Miller Property**, where the owner has exposed new veins for collectors. Recent visits by clubs and individuals have yielded many good crystals to around 20 centimeters.

Other Ontario CMB localities that are too numerous to mention have produced good apatite specimens, but a few more are worth noting. In Frontenac County, the **Bob's Lake Chlorapatite Occurrence** is notable as the only known occurrence of an apatite group mineral other than fluorapatite from Grenville province. And interestingly, this "chlorapatite" is monoclinic rather than hexagonal (Hounslow & Chao 1970) and is thus "chlorapatite-M," but is instead an unnamed

Open for Collecting
The Bear Lake Diggings, Haliburton County, Canada

Upper left: Welcome sign erected at the entrance to the property by the Bancroft & District Chamber of Commerce.

Top right: Weathered calcite vein exposed by collectors, approximately 1.5 meters wide.

Middle right: Ontario collector Alfred Newnham at the 2012 Gemboree, showing off the 34 cm crystal he found that spring.

Lower left: The calcite core of a weathered vein, showing apatite *in situ*, that was excavated in the fall of 2011

Below: Fluorapatite, 6 cm long crystal
George Thompson collection

All of the photos on this page are by Michael Bainbridge.

Upper left: Fluorapatite, 4.54 carats
Silver Crater Mine, Hastings County, Ontario
Robert Beckett collection, Michael Bainbridge photo

Upper right: Fluorapatite, 7.5 cm tall
Canoe Lake, Frontenac County, Ontario
Canadian Museum of Nature collection (cat. no. 51004; ex George Robinson) Michael Bainbridge photo

Lower left: Fluorapatite (repaired), 9.5 cm tall
Renfrew County, Ontario
Dustin Ziegler collection, Joe Budd photo

Lower right: Chlorapatite, 3 cm wide
Bob's Lake Chlorapatite Occurence, Frontenac County, Ontario
John Montgomery collection, Michael Bainbridge photo

polymorph of chlorapatite (see page 6). The best specimens are pink-white with chalky luster, up to 10 centimeters long and 2½ centimeters in diameter. Also in Frontenac County, the **Canoe Lake Roadcut** (aka **Desert Lake**, **Verona**, **Bob's Lake**) is noteworthy for producing deep blue apatite; the best is similar to material from Brazil and Russia.

Just west of Tory Hill, in Haliburton County, (on the same road that leads to the Bear Lake Diggings), is the **Gibsons East Occurrence** (aka **Titanite Hill**). With apatite crystals that tend to be closer in appearance to Liscombe material than to Bear Lake (although not generally of such consistently good quality), it has become a popular collecting destination. Near Wilberforce (also Haliburton County), the **Cardiff Uranium Mine** and **Richardson (Fission) Mine** have produced excellent, deep forest-green, transparent crystals to several centimeters in a banded cream-white calcite and purple fluorite matrix. And new claims for public collecting farther along Mumford Road will undoubtedly produce worthy specimens in the near future. At the nearby **Silver Crater Mine**, in Hastings County, beautiful deep green apatite crystals are found associated with the outstanding betafite group minerals for which the site is best known.

Several localities in the Perth area have produced good specimens, but few (if any) are still open to collecting. The **Silver Queen Mine** in Murphys Point Provincial Park, however, runs guided tours of the underground workings through the fair-weather season.

Québec

Probably the most prolific and well-known of all the Grenville apatite specimen producers is the erroneously named **Yates Uranium Mine** (aka **Otter Lake**), Huddersfield Township, Pontiac County (Pontiac RCM). The deposit was explored in the late 1800s for mica then fluorite and later, in the 1950s and 60s by J.M. Yates and others, for uranium. A handful of small but good examples of radioactive species, including allanite and thorianite, have been found, but

significant uranium mineralization has never been noted.

George Robinson and Frank Melanson worked the **Matte Zone** (one of six showings spread across eight, 100-acre lots) from 1978 to 1981. They recovered many exceptional apatite specimens, but others have collected large numbers of fine specimens over the years as well as this site has been a favorite field-trip destination for local mineral clubs since the mid-1960s. From 1997 to 2011, Darryl MacFarlane and partner Fred Sopchyshyn (who purchased the interest of the late Jerry Van Velthuizen) worked the deposit for specimens, and although good crystal potential still exists, extraction has become prohibitively difficult and expensive. Specimen production has thus ceased, and the site was reclaimed in 2011 (Darryl MacFarlane, personal communication).

The occurrence lies at or just beyond the northwestern limit of the CMB in southwestern Québec. The apatite crystals are typically far sharper than those found at most other CMB occurrences, and the site has produced numerous excellent specimens with some crystals exceeding 30 centimeters in length; in fact, crystals reaching 50 centimeters in length have been noted *in situ*. The apatite is associated with dark green pyroxene, white scapolite, microcline, and purple fluorite, all found embedded in orange calcite. At best, the apatite is a deep, dark transparent green but in certain areas of the deposit it may be brownish red. The single, large calcite vein-dike which has produced these specimens swells to 5 meters in width and has been exposed over a length of about 200 meters.

As with Ontario, there are innumerable good apatite localities in the CMB of Québec, and many excellent specimens have been found in the Pontiac-Gatineau-Buckingham regions. A 500-pound crystal from the Comet Mine was exhibited in Paris in 1878 and is now in the collection of the Canadian Museum of Nature, Ottawa. Apparently, it used to be much larger, but the ship's captain considered it "too bulky" and broke it into several pieces to make it easier to transport. (Van Velthuizen 1992). Noteworthy localities in this area include the **Seybold Mine**, which is known for unusual sky-blue apatite, sometimes with a dark blue or inky black outer zone; the **McGlashan Mine**, which has produced excellent crystals in shades of blue-green; and the **Comet Mine** near "Wilson's Corner." Crystals from all three of these localities tend to be largely opaque, but their attractive colors distinguish them from the common green crystals found in the popular Ontario localities. Although gaining access to collecting sites is becoming increasingly more difficult, collectors from Ottawa and Montreal still make productive trips to the area.

Outside of the Grenville province, white quartz veins cut strata of metamorphosed sediment in southern Québec. Many of these veins host a simple mineralogy, but large cavities of more complex Alpine-type mineralization are occasionally encountered, especially in a series of road cuts on Québec **Highway 55**, north of Sherbrooke. A cluster of six separate pockets confined to vertically dipping, gray-green possibly metavolcanic units are exposed on the western side of the southern road cut. Tabular pink fluorapatite crystals up to 1.4 centimeters occur here and other localities in the area. The apatite is associated with albite, anatase, and chamosite.

Right:
Fluorapatite, 5 cm wide
Collected by Brad Wilson
Little Nahanni Pegmatite Group,
Northwest Territories
Brad Wilson collection
Michael Bainbridge photo

Facing page, upper left:
Fluorapatite, 1.3 cm wide
Collected by Michel Picard and Gilles Haineault
Highway 55 roadcut north of Sherbrooke, Estrie Region, Québec
Canadian Museum of Nature collection (cat. no. 52663)
Michael Bainbridge photo

Facing page, right:
Fluorapatite, 10 cm tall
NOT repaired!
Collected by Brad Wilson
Yates Uranium Mine
Otter Lake, Québec
Brad Wilson collection
Michael Bainbridge photo

Facing page, lower left:
Fluorapatite, 1.36 carats
Faceted by Brad Wilson
Danford Lake, Estrie Region, Québec
Brad Wilson collection
Michael Bainbridge photo

THE NORTH

While the Grenville has been extensively studied, the Canadian Cordillera and the north in general, have not. Canada's north country is beautiful in spite of the hoards of black flies and mosquitoes, extreme weather conditions, and the fact that it is plunged into 24-hour darkness for approximately 6 months of the year. No wonder it is largely unexplored! And in fairness, the "North" cannot be characterized by geography, geology, or culture. But what the localities of the north do have in common (aside from the cold) is the select group of intrepid mineral seekers, who gravitate to them.

The Yukon covers almost 500,000 square kilometers and has a population of just over 31,000 people (almost identical to the population back in 1900, incidentally), two-thirds of whom are concentrated in the capital city of Whitehorse. Only a few key specimen producers, such as Emerald Lake and the Rapid Creek/Big Fish River area, have been identified. By comparison, the population of the Northwest Territories is 41,000, but the territory is more than twice the size of the Yukon. The **Little Nahanni Pegmatite Group** is perhaps the only marginally well-known producer of apatite so far found in the Cordillera of the Northwest Territories. Then there's Nunavut, with only 1.3 people per 100 square kilometers. Its 2,000,000 square kilometers make up 20 percent of Canada's landmass. The Central Medasedimentary Belt of Baffin Island (near Kimmirut, Nunavut) has mineralization similar to that of the Grenville area, but very few mineral specimens have yet been recovered, as the localities are all but inaccessible. No roads connect communities in the province; instead all are accessed by snowmobile (in winter) or air. Instead of people and infrastructure, Nunavut's pristine majestic mountains, glaciers, rivers, and areas of flat tundra are full of grizzly bears, caribou, mountain sheep, and muskoxen.

"**Rapid Creek**" in the Yukon's Dawson mining district covers a fairly large area and includes numerous localities each with its own most notable species, but all in a phosphate-rich ironstone. Although better known for rarer phosphates such as lazulite, whiteite group, gormanite-souzalite, and bobdownsite-whitlockite, this area (three sites in particular) has also yielded beautiful purple crystals of apatite. The finest specimens originate in an unnamed occurrence along an unnamed tributary of Rapid Creek that is approximately 1.7 kilometers south of another feeder known as "Crosscut Creek."

In 1983 George Robinson collected some very pale mauve apatite from **Kulan Camp**, and in 1984 Joel Grice and Bob Gault collected pale mauve crystals from **Stoneman Camp** on Grizzly Bear Creek. While the crystals at these sites are larger than those from the area south of Crosscut Creek above, their color is much paler. These apatite crystals are most commonly associated with lazulite, quartz, and siderite. The crystals tend to be fairly stubby and range in size from microscopic to nearly 2 centimeters long.

During the summer of 2012, a research team led by Kimberly Tait, Curator of Mineralogy at the Royal Ontario Museum and author (KT) of the present text, spent two weeks collecting in the **Big Fish River** area (about 13 kilometers east

Above: Fluorapatite with lazulite, quartz, and siderite, 5.5 cm wide
Rapid Creek, Yukon
Michael Bainbridge collection and photo

Below: Fluorapatite, 2.8 cm wide
Peterson Pocket, Big Fish River, Dawson district, Yukon
Royal Ontario Museum collection, Michael Bainbridge photo

of Rapid Creek) and also in the Dawson mining district. Well-crystallized arrojadite group minerals were found associated with quartz in a series of veins crosscutting the flat-lying sedimentary layers at the site. Small (3- to 7-millimeter-long), colorless apatite prisms are common. The apatite represents a new association for the locality.

Immediately below the bobdownsite-whitlockite locality, Ron Peterson of Queen's University in Kingston, Ontario, spotted a relatively large pocket in the base of a cliff. Standing in waist-deep, freezing-cold water, Royal Ontario Museum Collections Technician Ian Nicklin and collector Robert Beckett spent several hours recovering specimens from the "**Peterson Pocket**," which yielded the best apatite of the trip. The crystals are so unusually tabular that they were initially mistaken for whitlockite. As it turns out, each large "crystal" is actually multiple crystals composed of tiny prisms in parallel growth, with the overall appearance of a single larger crystal with hexagonal outlines. The crystals are opaque, milky-white to pale yellow (due to iron staining) and are up to 14 millimeters in diameter and 3 millimeters thick, sometimes tapering to less than 1 millimeter at the edge.

Some of the finest apatite crystals yet found in Canada originate from the **Scepter Claim**, high in the cliffs of a hanging valley above Emerald Lake, Mayo mining district, Yukon. Rod Tyson, Brad Wilson, John Gorham and Andrew Locock set up camp some 900 meters below a zone of miarolitic cavities on the claim, which is owned by Tyson. Working these cavities in the summers of 1994, 95, and 96, they recovered some spectacular apatite. Each day they hiked over a glacier and mounted a highly technical ascent to the collecting area. On one trip the team set up a portaledge hundreds of meters up the cliff face and right on top of a pocket. When government geologists flew in to inspect the mining activities (and the extreme campsite), they declined even to get out of the helicopter!

On one expedition, Wilson and Tyson completed a (now infamous) single-line rappel, 350 meters down a cliff to look at several very large pockets. Although two of these openings were large enough to accommodate a Volkswagen microbus, they were completely empty. But overall, their efforts paid off. One eroded miarolitic cavity at about the 2250-meter level produced several hundred loose apatite crystals including one that is about 3-by-3-by-3 centimeters (most were 0.5 centimeters in diameter and 1.5 to 2 centimeters in length). Only one matrix apatite piece was recovered from the site: a smoky quartz crystal with three apatite crystals adhering to its base found in another pocket. Apatite, smoky quartz, schorl, and scheelite are among the collectible species that have been found on the Scepter Claim. Broken crystals of apatite (and other gem minerals including scheelite) provided good rough for a number of faceted stones to around seven carats.

IN CONCLUSION

Nearly 200 years ago, Dr. Wilson made a series of discoveries in southern Ontario that led Canada to become a major producer of phosphate worldwide; today similar opportunity awaits in Canada's largely unexplored northern territories. But even as lucrative economic deposits or key specimen localities are found, the lack of infrastructure to access these areas makes systematic exploitation nearly impossible—for now. Canada is truly a vast country, with much of the story of apatite still waiting to be written.

ACKNOWLEDGEMENTS: *The authors extend their thanks to Robert Beckett, Gilles Haineault, Darryl MacFarlane, Frank Melanson, Ian Nicklin, Michel Picard, George Robinson, George Thompson, Rod Tyson, and Brad Wilson, all of whom provided information for this article.*

Top: Andrew Locock in the mouth of a 5-meter-deep, tube-like vug at the Scepter Claim at Emerald Lake in the Yukon. A team including Brad Wilson worked the pocket as far as the permafrost in 1994. They returned in 1996 to find that the permafrost had retreated, and they finished collecting the pocket. Both the apatite crystal and gemstone pictured above, along with most of the best material, was recovered from that pocket in that year
Brad Wilson photo

Lower right: Fluorapatite, 2.4 cm tall crystal, 7.49 carat cut stone
Scepter Claim, Emerald Lake, Yukon
Brad Wilson collection, Michael Bainbridge photo

Left: Fluorapatite, 4 cm tall
Kulan Camp, Rapid Creek, Yukon
Canadian Museum of Nature collection
(cat. no. 55544; ex Bill Pinch, Van Velthuizen)
Michael Bainbridge photo

Mexican Apatite

Peter K. M. Megaw and Mark Barton

Topping the list of Mexico's "apatite" specimens are of course the sharp yellow-green gemmy prismatic fluorapatites ("asparagus stone") from Cerro de Mercado in Durango. Known to the mineralogical and gemological world since the 1850s, these crystals were once so abundant that they were sold by the bucketful and were unhesitatingly adopted worldwide as laboratory standards. When the Cerro de Mercado metallurgical plant converted from hand-sorting to magnetic separation in the 1990s, this abundant source for apatite specimens was eliminated, although large gemmy crystals continue to trickle out in mine-workers' lunchboxes.

Other Mexican apatite localities include the prolific Guadalupana and other tourmaline "mines" in the Sierra el Chivato area, Santa Cruz, Sonora, which produce substantially fewer and less attractive colorless apatite crystals that occur with dravite-uvite (often mislabeled schorl). And the famous "coconut" geodes from Las Choyas, Chihuahua, contain microscopic carbonate-rich apatite crystals (Finkleman and others 1972, and Finkleman 1974).

Two localities in Mexico are important producers of apatite pseudomorphs. Stunning blue pseudomorphs of turquoise after stout prismatic apatite crystals up to 15 centimeters long have come intermittently from the Cumobabi and Caridad Mines in Sonora, and pale tan hydroxylapatite pseudomorphs after several unidentified but well-crystallized species come from the Astillero Mine in Mapimi Municipality. The Astillero Mine is 25 kilometers from the famous Ojuela Mine, but specimens are commonly misattributed to the latter. In fact misattributed specimens are widely distributed in older collections with a wide range of amusing misidentifications of both the original and replacing species.

CERRO DE MERCADO

Cerro de Mercado, also incorrectly known as "Iron Mountain," "Cerro Mercado," or "Cerro del Mercado" (Zubiria y Campa 1922a), is today almost completely surrounded by the northern outskirts of the city of Victoria de Durango, capital of the Mexican state of Durango. This legendary "mountain of iron" is an east-west oriented triangular-shaped ridge approximately 2 kilometers long, tapering eastward from its 1.5-kilometer-wide western base. The rugged hill with large open pit and dumps rises abruptly 100 meters to 150 meters above the surrounding city neighborhoods. Cerro de Mercado is composed of a series of massive iron-oxide bodies occurring within a sequence of Tertiary rhyolitic volcanic rocks; it is renowned for large octahedral "martite" (hematite after magnetite pseudomorph) specimens and gemmy fluorapatite crystals that occur in fractures and breccia bodies cutting the massive iron ore.

History

Cerro de Mercado takes its name from Don Ginés Vázquez del Mercado who was sent in 1552 to conquer the Guadiana Valley, site of the modern city of Victoria de Durango (Tello 1653; Frejes 1839; Garcia 1895; and Zubiria y Campa 1924). The expedition was also tasked with investigating persistent rumors of a marvelous mountain of gossan with native gold

Above: *Sobre el Cerro de Mercado*, "Above the Cerro de Mercado"; Ivan Rumata photo

Facing page: Fluorapatite, 6.4 cm tall
Cerro de Mercado, Durango, Mexico
Peter Megaw collection, Jeff Scovil photo

hydrothermal origin (Rangel 1902), triggering a debate that continues into the twenty-first century (Labarthe and others, 1987; Lyons 1988; Barton and Johnson, 1996 and 2000).

Mineralogical History

Del Rio (1835) described octahedral martite from Cerro de Mercado in his famed treatise on the mineralogy of Mexico, but Weidner's (1858) description of prismatic gemmy crystals of "fenaquita" appears to be the first description of fluorapatite. De Landero (1888) and Aguilera-Serrano (1898) also mentioned what we now know is fluorapatite as did Cohen (1893), Rangel (1902), Farrington (1904), and Quintero (1923). Interestingly, Silliman (1882) focused exclusively on the martite. Foshag (1929) published what remains the most complete descriptive mineralogy of Cerro de Mercado, including fluorapatite. He describes variations shown between different areas within the deposit. The older hobby literature is well sprinkled with mentions of fluorapatite (Sinkankis 1959 and 1976) with Johnson (1963 and 1965) giving detailed descriptions and collecting advice. More recent mineralogical articles include Megaw and Barton (1999) and Jurgeit (2001).

Cerro de Mercado fluorapatite assumed a strong scientific/academic identity in the late 1960s and 70s as it was adopted as a phosphate standard by the United States Geological Survey's spectrochemical lab (Young and others 1969); used as a Fission Track Dating annealing standard (Naeser and Fleischer 1975); and was used as a reference material in thermoluminescence studies (Vaz 1980). Young and others (1969) presented detailed data on the chemical, physical, optical, and x-ray characteristics of the fluorapatite, including trace element substitutions in the crystal lattice.

and silver littering the surface. The mountain was quickly revealed to be purely iron and del Mercado's dreams of unimaginable wealth were dashed. Despondent, he headed home toward Guadalajara, but his party was attacked and he perished before reaching his destination. His name, however, became attached to the mountain. Some historians maintain that the mountain was named in honor of his giving his life in the quest for greater riches and glory for the Spanish crown while others claim that it derides his delusions of grandeur (Halse 1907).

The first organized mining/smelting at Cerro de Mercado began in 1828 and consisted simply of picking up bits of iron-ore talus and hauling them to a crude foundry (Ward 1829). Operations became more formal in 1834 when a blast furnace was built to process ore extracted from the earliest excavations on the hill (Escudero 1849). Mining and iron smelting expanded incrementally over the next 150 years to reach the current scale of 40,000 tons per month (Chrustschoff 1878; Iron Mountain Company 1882; Ramirez 1884; Witherbee 1902; Terrones-Benitez 1944; Mundo Minero 2011).

Since the sixteenth century, explorers, miners, opportunists, and geologists have investigated the nature of the iron mountain. Early ideas include Vázquez del Mercado's notion of an enormous gossan with untold wealth beneath. Baron Alexander von Humboldt, who never even visited Durango, is erroneously cited as the source of a popular but incorrect early nineteenth century theory that the deposit was a huge meteorite (Weidner 1858, 1877; Zubiria y Campa 1922b). This misinformation may stem from conflation of von Humboldt's description of an iron meteorite presumed to be from the Durango area and his second-hand mention of Cerro de Mercado as a mountain of magnetic iron ore (Burkhart 1858).

Modern geologic theories began to take shape during the nineteenth century. Weidner (1858) gave the first detailed description of the hill and invoked a volcanic origin after an exhaustive argument on why it could not be a meteorite. Weidner was followed by a string of geological and mineralogical visitors, most of whom argued for an igneous origin (Birkenbine 1885; Farrington 1904; Salazar-Salinas and others, 1923; and Flores 1950). Others advocated a

Geology

The city of Durango and the Cerro de Mercado lie within the Chupaderos Caldera, a large rhyolitic volcanic center that is part of one of the world's largest piles of felsic volcanic rocks—the middle Tertiary Sierra Madre Occidental volcanic province (Swanson and others 1978; McDowell and Clabaugh 1979; Lyons 1988). Chupaderos Caldera volcanism began with large volume ash-flow tuffs and culminated with rhyolitic domes, intrusions, and airfall tuffs that host the iron-rich bodies at Cerro de Mercado and elsewhere in the district (Lyons 1988). Some of the rhyolite domes contain wood tin (botryoidal cassiterite) and may be the sources of the beryl and phenakite that was erroneously attributed to Cerro de Mercado in older literature (Megaw and Barton 1999).

The iron deposits of Cerro de Mercado form a series of massive lenses and breccias that are broadly conformable with the surrounding rhyolitic rocks, although in places they clearly cut and are cut by silicate-altered rocks (Foshag 1929; Barton and Johnson 1996). The original mass extended more than 800 meters in a north-south direction, reaching a thickness of well

Right:
Fluorapatite, 6 cm tall
Cerro de Mercado, Durango
Mexico
Peter Megaw collection
Jeff Scovil photo

Facing page:
General location of the Cerro de Mercado Mine, Durango
Mexico
William Besse cartographer

over 100 meters. Lyons (1988) describes the oxide-rich rocks as consisting of basal, layered, variably brecciated martite, overlain by a "sandy" martite unit, and capped by a mixed unit of blocky quartz latite with a martite matrix. A late cross-cutting zone of iron oxides contains both massive and brecciated zones. The deposit is surrounded by volcanic rocks that have been variably converted to mixtures of pyroxene, magnetite, hematite, quartz, and clays (Foshag 1929; Barton and Johnson 1996). Overall, Cerro de Mercado shows many features indicative of a magmatic origin (Lyons, 1988) but also shows the strong hydrothermal effects (Barton and Johnson, 1996) that led to the kind of conflicting genetic interpretations described by Martin and Rakovan (see page 21).

Fluorapatite is part of at least three stages in the paragenesis (Megaw and Barton 1999). Early brownish fluorapatite forms both stubby crystals and dendritic fans of elongate crystals ranging from 0.5 to 3 centimeters across and up to 30 centimeters long that are locally accompanied by iron-rich clinopyroxene and frozen in iron ore. This early fluorapatite is commonly partly to completely replaced by fine-grained quartz. The next stage is dominated by hydrothermal quartz and the famous yellow-green fluorapatite crystals, which occur singly or as vein fillings and breccias cemented by minor sepiolite, hematite, gypsum, and calcite. A final generation of fluorapatite (possibly carbonate-fluorapatite or hydroxylapatite) appears as fine-grained druses of elongate needles or prisms on martite and on earlier yellow-green fluorapatite (Foshag 1929). This last generation may be contemporaneous with fine-grained crystalline goethite that grows on broken martite crystals. Chalcedony overgrowths are common on all stages of fluorapatite and hematite, indicating that late-stage vapor-phase silica mineralization was extensive during the waning phases of deposition. The multiple phases of fluorapatite growth, dissolution and regrowth indicate that it was readily remobilized and reprecipitated throughout the history of the deposit (Megaw and Barton 1999).

Cerro de Mercado Fluorapatite

Fluorapatite ($Ca_5(PO_4)_3F$) is the dominant and best known apatite species from Cerro de Mercado, occurring as superb gemmy greenish-yellow to golden crystals to 10 centimeters in length. It is also commonly cut into fine semiprecious gemstones called "asparagus-stone" (*esparraguina*). The crystals show dominant first-order prisms with minor development of the second-order prisms and sharp unit pyramid terminations (Foshag 1929). Etching and regrowth locally make for more modified terminations (Megaw and Barton 1999). Rangel (1902) noted that the best fluorapatite

crystals came from the *Cueva de los Biriles* ("Cave of the Beryls") area, where they occurred as vein fillings and masses of well-formed crystals. Unfortunately this cave was mined away years ago, but it was probably similar to the vuggy quartz-cemented breccia bodies filled with fluorapatite seen today in the deeper parts of the pit. Single, loose fluorapatite crystals were in abundant supply during the pre-1985 mining period, as they were systematically hand-picked off the conveyor belts to reduce the phosphorus content of the iron ore. Unesteemed by the mining company, they were sold at the mine offices for US$5 per bucketful. Miners, who knew that a few good "amarillos" made for a tidy self-awarded bonus, made undamaged matrix specimens available for sale. Today's higher mine throughput and mechanized magnetic sorting means most crystals are crushed to dust, and only those few pocketed by the miners survive.

Carbonate-hydroxylapatite, $Ca_5(PO_4,CO_3)_3(OH)$, formerly known as "dahllite" was reported by Foshag (1928) as hexagonal plates and prisms, to 1 millimeter across, coating martite vugs. No analytical work has been done on this material, but given the preponderance of fluorapatite, Foshag may have actually been referring to carbonate-fluorapatite, $Ca_5(PO_4,CO_3)_3F$ (Megaw and Barton 1999).

Cleaning

Cerro de Mercado fluorapatite crystals are often coated with unsightly calcite, clays, gypsum, iron oxides and/or chalcedony (Megaw and Barton 1999). Because of the relative softness of fluorapatite, abrasives or metal tools should not be used to remove these coatings. In most cases, calcite, clay, and gypsum coatings may be carefully scaled from fluorapatite with a fingernail or bamboo skewer. However, many of the crystals are crazed and healed with calcite so when removing any coating, care must be taken on approaching the tips of the crystals to avoid putting pressure on the calcite cemented fractures and taking the point of the termination off with the coating. Removing iron oxides and chalcedony is a greater gamble and challenge because, in addition to the healed fractures, the underlying fluorapatite often has a matte luster caused by the fluids that deposited these coatings, and iron oxides commonly penetrate deep into the crystals along fractures. Fluorapatite is readily etched by acids, but some commercial, non-acid, iron-oxide removing preparations do an adequate job of removing iron stains. Soaking in frequently changed baths of pure distilled water is a slow process, but one that can be successful. In any case, heating specimens to more than 100 degrees Celsius should be avoided as the crystals may lose their color (Foshag 1929). Chalcedony coatings are difficult to remove, and the underlying surfaces are almost always deeply etched, so efforts should often focus on cleaning the chalcedony rather than removing it.

APATITE PSEUDOMORPHS

Blue and blue-green turquoise pseudomorphs after stout hexagonal apatite prisms have intermittently come from the **Cumobabi** and **La Caridad** mines in northern Sonora. These somewhat crude crystals range up to 15 centimeters in length. Terminations are low, simple pyramids truncated by a pinacoid. Many of these pieces are intergrown with coarse fans of bladed biotite up to 6 centimeters across and remnants of crumbly euhedral quartz crystals. Very similar, but unoxidized versions of this assemblage have been found at **Mina Maria** and in the **La Colorada** and other breccia pipes in the Cananea District (Wodzicki 1995). These breccia pipes are "rooted" in volatile-rich cupolas that developed on the tops of the principal porphyry copper-related intrusions. Crystallization conditions can be considered transitional between magmatic and hydrothermal; they could be called "hydrothermal pegmatites." The district-wide relationships suggest that primary apatite was converted to turquoise either during late copper mineralization or supergene leaching of the deposit.

These distinctive crystals have been labeled as coming from "Cananea," "La Caridad," "Nacozari," "Baviacora," "Cumobabi," and even "Baja California." An historic source for these specimens maintains that the older, greener specimens (pre-1990s) came from Cumobabi and the subsequent, bluer examples came from La Caridad.

The **Astillero Mine** was a small operation in the Sierra el Rosario, roughly 20 kilometers southwest of the famous Ojuela Mine in the state of Durango. Both are in the Municipio of Mapimi, and most specimen material from Astillero has been sold as coming from the Ojuela Mine. The Astillero Mine exploited a series of small pods of oxidized lead-zinc-silver replacement mineralization emplaced in limestones along the contact of a small stock. It is currently being explored by Grupo Mexico.

The Astillero Mine produced a moderate number of pale tan to off-white pseudomorphs of hydroxylapatite replacing at least three unidentified species. Two of the pseudomorphed species had elongated, bladed habits (one with strong longitudinal striations), and the third had more equant, rounded blades. Stibnite, hemimorphite, anhydrite, and gypsum have been suggested as the former and barite or gypsum for the latter. Given the habits, striations, lack of antimony in the pseudomorphs, and common chemistry, anhydrite and gypsum are reasonable candidates.

Right:
Hydroxylapatite pseudomorph possibly after anhydrite or gypsum
5.6 cm high
Astillero Mine, Mapimi, Durango, Mexico
Peter Megaw collection
Jeff Scovil photo

Below:
Turquoise pseudomorph after Apatite, 5.5 cm tall
La Caridad, Sonora, Mexico
Peter Megaw collection
Jeff Scovil photo

Facing page:
"Asparagus-stone," faceted fluorapatite, 24.61 carats
Cerro de Mercado, Durango, Mexico
John Lucking collection
Jeff Scovil photo

Less reasonable are the historical misidentifications of what recent x-ray and microprobe analyses have shown to be hydroxylapatite (Marcus Origlieri, unpublished data). "Forbesite," a discredited species now known to be a mixture of cobaltian annabergite and arsenolite, is the most common misidentification, with stibiconite, mimetite, and "wilkeite" (another discredited species now recognized as siliceous sulfatian apatite) adorning other labels. At least the latter two have apatite group structures!

CONCLUSION

Cerro de Mercado fluorapatite continues to trickle from the mine into the specimen market, but the other apatite localities in Mexico are largely inactive so specimens of "apatite" from them have to be sought from dealer stock and old collections. Apatites have been intermittently found in small amounts from some of Mexico's large skarn and carbonate replacement deposits, so potential exists for new finds to emerge from new localities. Whether these can challenge the classics from Cerro de Mercado is unknown.

ACKNOWLEDGEMENTS: *The authors are grateful to the management of the Cerro de Mercado Mine for the opportunity to map and study their deposit. Longtime friend and colleague James Lyons is due thanks for years of stimulating discussion on Cerro de Mercado and other places in the Sierra Madre. Marcus Origlieri kindly analyzed the Astillero material. He would also like to thank IMDEX Inc. for the use of MEXBIB 3.3 © to generate the bibliography.*

Brazil's Great Apatite Localities

by Luiz Menezes and Mario Chaves

For decades Brazil has been a major source for outstanding apatite crystals, which are found in the country in a variety of colors and habits. Most of Brazil's apatite comes from granitic pegmatites, but other types of deposits also produce collectible specimens. Brazil's most important apatite occurrences are in the Oriental Pegmatite Province of Minas Gerais and in the granitic pegmatites in the Borborema Pegmatite Province (which covers parts of the northern states of Rio Grande do Norte and Paraíba).

Map: Brazil's Major Apatite Deposits
William W. Besse cartography

Facing page: Fluorapatite on albite, 2.9 cm wide
Jonas Mine, Conselheiro Pena, Minas Gerais, Brazil
Luiz Menezes collection, Marcelo Lerner photo

ORIENTAL PEGMATITE PROVINCE

The granitic pegmatites of the Oriental Pegmatite Province of Minas Gerais are Brazil's most important source of mineral specimens and gemstones. This well-known province crosses three states: northeastern Minas Gerais, southeastern Bahia, and the western portion of Espírito Santo. Several hundred pegmatites have been discovered in the province and many more are yet to be found. The pegmatites formed between 610 and 480 million years ago as a result of intense magmatism related to the Orogen Araçuaí, a geologic belt which resulted from the collision of the West African and the South American tectonic plates.

The province runs north to south for approximately 800 kilometers and east to west for about 150 kilometers. Individual pegmatites are distributed across several districts, each of which is associated with a large granite batholith. While the mineralogies of the pegmatites within a given district may vary, the bulk of a district's pegmatites are either elbaite-bearing, aquamarine-bearing (those containing euhedral, blue beryl crystals), or neither. Pink, purple, or blue apatite crystals tend to occur in the province's elbaite-bearing pegmatites while yellow or green apatite is found in those that are aquamarine-bearing; in the others, anything goes.

Elbaite-Bearing Pegmatites

Elbaite-bearing "pegs" are generally 10 to more than 30 meters thick, tend to be several hundreds of meters to a few kilometers away from their granite sources, and often show complex zoning. While a contact zone is only sometimes present, a wall zone, outer intermediate zone, inner intermediate zone, quartz core and albite replacement bodies are all common components of these pegmatites. Apatite occurs as irregular grains in the intermediate zones, but the best specimens are found in the albite replacement zones, which are located on the inner intermediate zone surrounding the quartz core or on the borders of that core. The apatite found in this type of pegmatite tends to be pink, purple, or blue. Yellow, green, or brown apatite has, as yet, not been observed in these pegmatites.

In terms of quantity, the **Golconda district** in Governador Valadares, Minas Gerais, has been Brazil's primary source for pegmatitic apatite. The most important producers are Lavra da Golconda II, Lavra do Faria, Lavra Zé Pedra and Lavra do Escondido. *Lavra* is Portuguese for "small mine."

In the 1960s, when the "**Golconda Mine**" became known around the world as a major source of gem-grade green and blue elbaite, there were two important mines known as "Golconda": one, which for clarity we shall call "**Lavra da**

Above: Fluorapatite and quartz, 11.5 cm wide
Lavra do Faria, Governador Valadares, Minas Gerais, Brazil
Gail and Jim Spann collection, Jeff Scovil photo

Right: The Golconda (II) Mine
Guido Steger photo, 1985

Below: Fluorapatite on muscovite with bertrandite and albite
Field of view is about 9 cm wide
Lavra do Faria, Governador Valadares, Minas Gerais, Brazil
Sam Yuen collection, Jeff Scovil photo

Facing page, upper right: Fluorapatite and elbaite, 7 cm wide
Lavra do Faria, Governador Valadares, Minas Gerais, Brazil
Scott Rudolph collection, Jeff Scovil photo

Facing page, lower right: Fluorapatite on albite, 8 cm tall
Golconda II Mine, Governador Valadares, Minas Gerais, Brazil
Marty Zinn collection, Jeff Scovil photo

Facing page, lower left: Fluorapatite on albite, 5.2 cm tall
Golconda II Mine, Governador Valadares, Minas Gerais, Brazil
Luiz Menezes collection, Marcelo Lerner photo

Golconda I," was the source of the gem-elbaite. There is no record of significant apatite finds from this mine, which was exhausted in the late 1960s and has never re-opened. At about the same time and only one kilometer from the first, a second Golconda Mine was producing industrial quantities of muscovite, feldspar, beryl, and eventually fine quartz crystals, small amounts of elbaite, fluorapatite, and bertrandite. We will call this second mine, which has produced a few important pockets of apatite, "**Lavra da Golconda II**." The best of these pockets, opened in 1998, produced a lot of pink fluorapatite crystals (up to 3 centimeters), most of which were on albite (photo lower right). In 1993 an outstanding small pocket contained long, prismatic, transparent pale purple fluorapatite crystals up to 8 centimeters long (photo lower left).

Lavra do Faria has consistently produced small to medium-sized pockets containing blue to purple fluorapatite crystals, which generally reach lengths of up to 5 millimeters (rarely 1 centimeter). The crystals can be blue or purple and are normally prismatic and terminated with basal pinacoids and short pyramidal faces. The matrix consists of corroded microcline crystals covered with platy albite (variety cleavelandite), colorless quartz, yellow to gray muscovite, dark green elbaite, schorl, and millimeter-sized, colorless, flattened bertrandite crystals with a pearly luster on the basal faces. The latter species is a characteristic association not only of apatite from Faria but also from a number of other pegmatites in the Golconda district (e.g., Golconda II and Zé Pedra).

In just the last two years, this mine produced three remarkable pockets of fluorapatite. In the first of these, blue apatite crystals formed on pseudomorphs of cookeite after spodumene that are on large colorless quartz crystals. The other two pockets contained tiny blue fluorapatite crystals attached

Sapo Mine, Conselheiro Pena
Minas Gerais, Brazil
Left:
Fluorapatite "scorpion tails"
to 12 cm tall
Collector's Edge, Inc., specimen
Jeff Scovil photo
Facing page, left:
Fluorapatite, 11.3 cm tall
A pale gray, hollow "scorpion tails" growing over solid scorpion tail-like crystals
Luiz Menezes collection
Jeff Scovil photo
Facing page, lower right:
Road into the Sapo Mine
Guido Steger photo, 2003
Facing page, upper right
Hydroxylapatite, 4.9 cm wide
Luiz Menezes collection
Jeff Scovil photo

to single quartz crystals or to clusters of irregularly-shaped, tabular quartz crystals, which measure up to 35 centimeters. The pockets also contained apatite on microcline, albite, and muscovite and millimeter-sized, colorless to white bertrandite crystals (photo page 56, lower left).

In the past, **Lavra Zé Pedra** produced several outstanding pockets of schorl, and since 2007, the mine has also produced several small lots of intensely purple or blue fluorapatite crystals, some of which reach 3 centimeters in length. The crystals are often prismatic, with striations parallel to their *c*-axes and simple basal terminations. Most of the specimens consist of single crystals or clusters of crystals. They were found loose in the pockets—matrix specimens are not commonly found at this locality. In 2000 a single pocket, found at **Lavra do Escondido**, produced a few hundred flattened, light pink fluorapatite crystals, some as long as 6 centimeters.

Elbaite-pegmatite districts outside of Golconda have also produced fine apatite crystals. In **Conselheiro Pena**, located about 80 kilometers southwest of Governador Valadares, the **Jonas Mine** was made famous by the 1978 find of gemmy red elbaite crystals, some exceeding lengths of 100 centimeters. Intensive subsequent work has yielded several small to medium size pockets of tourmaline (most showing dark green color), and in late 1980, one small but outstanding cavity containing fine, clear, dark pink apatite crystals on albite. The best specimen from this find was a perfect thumbnail (pictured on page 55).

One of the authors (LM, who was at the time working as a mining engineer) purchased this lot of apatite in January 1981, just before making his very first trip to the Tucson show. He sold almost all of the specimens at the show but kept that killer thumbnail in his personal collection. That first success helped him to convince his wife to support him in starting a part time mineral specimen business (thus began author Luiz Menezes' illustrious career as a full-time mineral dealer).

Also in Conselheiro Pena, the **Sapo Mine** has, since 1995, produced a great number of top quality, large quartz crystals (colorless, smoky, and citrine) as well as several pockets of gem elbaite, and many cavities containing blue, purple, or pink fluorapatite crystals. Two of those Sapo apatite pockets were spectacular. The first came in June 2005 after a huge cavity that produced more than 100 tons of large quartz crystals and giant microcline crystals had been excavated. A small vug in the vicinity of the emptied quartz cavity was uncovered. The vug contained a large number of broken microcline crystals associated with hydroxylapatite. The latter crystals were up to 5 centimeters in diameter and probably make up the largest single find of pegmatitic hydroxylapatite known from any locality anywhere (hydroxylapatite is relatively uncommon in pegmatites). Most of the crystals are zoned, containing a pale yellow core of fluorapatite and a wide green outer rim of hydroxylapatite (photo upper right).

In June 2007 another apatite-bearing pocket was found at Sapo; this one yielding a large number of fluorapatite crystals, some up to 8 centimeters in diameter. The pocket measured 4.5 by 2 by 2 meters and contained sharp, bright, flattened, dark green fluorapatite crystals. Many of the crystals formed atop albite and are associated with dark gray muscovite. The pocket produced several hundred top quality matrix pieces, varying in size from thumbnail (2.5 centimeters) to more than 50 centimeters long. Some of the muscovite clusters reach widths exceeding one meter, although many of the specimens were ultimately trimmed down to improve their marketability.

A number of clusters of opaque green, short prismatic stacks of apatite crystals, resembling scorpion tails, were also found in this spectacular pocket. The two best are about 18 centimeters long. One is a group of three "scorpion-tails," which diverge to form three branches—a tree-like formation. The other (pictured on the facing page) consists of three

Left:
Bicolor fluorapatite on albite, 6.2 cm tall
Pederneira Mine, São José da Safira
Minas Gerais, Brazil
Luiz Menezes collection
Marcelo Lerner photo

Facing page, top:
Fluorapatite on albite, 3.8 cm wide
Vista do Rio Mine, Conselheiro Pena
Minas Gerais, Brazil
Jerry Rosenthal collection, Jeff Scovil photo

Facing page, middle:
Fluorapatite and lepidolite on hollow elbaite
3.5 mm long
Pederneira Mine, São José da Safira
Minas Gerais, Brazil
Luiz Menezes collection
Marcelo Lerner photo

Facing page, bottom:
Apatite on albite, 1.5 cm
Pederneira Mine, São José da Safira
Minas Gerais, Brazil
Julio Landmann collection
Marcelo Lerner photo

"scorpion tails" (the largest is 12 centimeters long) on white albite matrix.

The pocket further contained a number of 5- to 8-centimeter-long clusters of medium to dark-green scorpion-tail-clusters of fluorapatite crystals, although these crystals have hollow centers. The hollow cores do not appear to have formed through the growth of the apatite over another mineral (which would have presumably later dissolved), as crystal faces are present on the surfaces of the hollow centers, suggesting a "hopper-type" formation. Perhaps these crystals grew quite rapidly such that the fluorapatite molecules were preferentially deposited on the peripheries of the crystals, leaving an empty center. The best of these specimens consist of five nearly parallel scorpion tails, the largest of which is 19 centimeters long. Pale beige examples of similar specimens were likewise recovered. The largest beige cluster, dubbed the "apatite snake," is a whopping 28 centimeters long!

Equant apatite crystals up to 15 by 15 centimeters were also found in this remarkable pocket albeit in lesser quantities than were the other colors and habits of apatite. Smaller crystals had developed in parallel growth with the primary crystals, which have solid cores and formed in colors grading from pale to dark green. The dark green color of these crystals is a result, not of chemical impurities, but of chlorite inclusions. Dark green chlorite was also abundant inside this pocket, where it partly coated the albite, muscovite, and apatite.

Finally, the **Vista do Rio Mine**, located on the shore of the Doce River, 5 kilometers from Conselheiro Pena, produced several small pockets containing flattened pink apatite crystals on albite and associated with blue elbaite, pink beryl ("morganite"), albite, quartz, and schorl.

In the São José da Safira pegmatite district the famous **Pederneira Mine** is the world's primary source of tourmaline specimens. The mine exploits two pegmatites, the lower of which was mined from the early 1970s until 1995. The lower peg produced outstanding elbaite crystals showing green, red, or blue, with many specimens exhibiting multiple color zones. It also hosted several fine small pockets of fluorapatite, the most important of which was exposed in 1994. The pocket yielded bicolor (blue and purple), prismatic, striated apatite crystals with basal terminations that are perched on pale yellow microcline crystals and are associated with small clusters of white albite (photo above).

A small pocket collected in 1995 from the lower pegmatite netted green elbaite crystals, the cores of which had corroded away. Inside the hollow crystals, tiny purple fluorapatite crystals had formed (facing page, middle right). Pretty pink crystals on albite were also found in the same period.

The mine was closed from 1995 until 2000, when the second pegmatite was found about 40 meters above and running parallel to the first. Since its discovery, this second pegmatite has produced an incredible number of outstanding elbaite specimens but almost no apatite. In a few pockets, green elbaite was associated with twinned crystals of hydroxylherderite, a species that was absent from the lower pegmatite thus indicating an interesting division of the secondary phosphates at Pederneira: fluorapatite occurred in the lower vein and hydroxylherderite in the upper.

In the same district, the **Cruzeiro Mine** (São José da Safira, Minas Gerais) is likely the world's richest tourmaline mine in that it has probably generated more tourmaline gem rough than any other mine in the world. The most prized of this material is the gemmy pink "rubellite" vareity of elbaite seen in portions of some of the elbaite crystals. The Cruzeiro Mine exploits at least three distinct pegmatites, all of which are deeply weathered—tourmaline is almost never found on matrix at this mine. The mine has occasionally yielded pink to purple fluorapatite crystals. In 1986 miners excavated a small pocket containing clear quartz crystals and green elbaite crystals with small, flattened light purple fluorapatite crystals on them.

There are no records of important apatite finds in either the Malacacheta district, where the Santa Rosa Mine is located, or the Itinga/Coronel Murta/Virgem da Lapa district, which encompasses numerous pegmatites including Barra do Salinas, Xanda, Limociro, Morro Redondo, Urubú, Baixão, Pirineus, Poço D'Anta, and Ouro Fino.

Aquamarine-Bearing Pegmatites

Cavities containing sharp, prismatic aquamarine crystals are found in small to medium-sized (rarely exceeding 10 meters in thickness) pegmatites in the Oriental Pegmatite Province. These pegmatites are not as well developed as the elbaite-bearing pegmatites and even though they are located close to or even inside the granite batholiths. The fluorapatite crystals found in these pegs tend to be yellow, green, or brown.

In 1989 the **José Pinto Mine** (Aldeia district, Conselheiro Pena, Minas Gerais) produced a large quantity of dark green fluorapatite crystals on "rosettes" of reddish brown muscovite. Individual fluorapatite crystals are almost equant and measure up to 3 centimeters while the muscovite rosettes can reach widths of up to 8 centimeters. The occurrence yielded more than one ton of muscovite clusters, although most were not associated with apatite.

The *município* (the rough equivalent of a county) of **Medina**, also in the state of Minas Gerais, is home to a large granite batholith with an associated series of small aquamarine-bearing pegmatites. The Serrinha Mine and Fazenda (meaning "farm") Campolina are among the operations that exploit these pegmatites, and they are the município's most important apatite sources. But a handful of unnamed pegmatites in Medina have also hosted small pockets, some of which contained fluorapatite crystals on albite matrix.

At **Serrinha** (which in 1997 produced an outstanding pocket of pencil-like gem aquamarine crystals up to 30 centimeters long), it is common to find small pockets of dark green apatite on an albite-microcline-muscovite matrix. In October 2005, a cavity was discovered that contained a doubly

terminated green-gray fluorapatite crystal measuring 17 by 11 centimeters. It is nicely positioned on a cluster of albite, which measures 28-by-24 centimeters. This is probably the largest fluorapatite crystal ever found in a Brazilian pegmatite.

In 1985, **Fazenda Campolina** yielded a cavity containing clusters of transparent, yellow fluorapatite crystals that are stacked one on top of the other such that the shapes of the clusters, like those from the Sapo Mine, are reminiscent of scorpion tails (although the Medina crystals are not hollow).

Numerous granite batholiths and associated pegmatites pepper the municípios of **Padre Paraíso**, **Catugi**, **Caraí**, and **Teófilo Otoni** in Minas Gerais. These pegmatites have variously been mined for aquamarine, heliodor, and topaz, and many have delivered small quantities of yellow or green fluorapatite, mostly as small centimeter-sized single crystals.

Several pegmatites in the state of Espírito Santo have been mined for aquamarine, clear colorless topaz, and more recently, quartz crystals. The latter shows zones of amethyst and smoky with hematite inclusions and is known as "super-seven" on the metaphysical market. The most important pegmatite-producing municípios in the state are **Fundão**, **Santa Maria do Jetibá**, **Itaguassú**, **Itarana**, **Santa Tereza**, **Colatina**, and **Mimoso do Sul**. Yellow, green, or brown fluorapatite crystals have been occasionally found in a number of these pegmatites. A recent discovery at Itarana produced nice, sharp, yellow apatite crystals with acicular inclusions oriented parallel to the *c*-axis such that they create a "cat's-eye" effect.

Other Types of Pegmatites

Many pegmatites in the Oriental Province have produced neither gem elbaite nor gem aquamarine but have yielded interesting apatite specimens. The **Urucum Mine** (Galiléia, Minas Gerais), for example is embedded in granite and should thus theoretically produce aquamarine, but it has instead been an important source for pink spodumene (variety kunzite) and of pink beryl (variety morganite). In fact, in August 1998, a giant cavity containing more than 100 kilograms of top-quality gem-grade kunzite crystals and more than 100 tons of quarts clusters (the largest crystal weighed a whopping 6 tons) was excavated. Several fine fluorapatite crystals were also collected from the 15-by-10-by-10-meter cavity. Some took shape as flattened, pale to medium purple crystals (one of the best, pictured above left, shows blue-purple zoning). Others formed large "skeletal" clusters of blue-gray crystals that measure up to 11.5 centimeters and are associated with etched spessartine crystals.

The **Navegadora Mine** (Penha do Norte district,

Right:
Fluorapatite, 8.8 cm tall
Teófilo Otoni, Minas Gerais
Brazil
Western Mineral specimen
Jeff Scovil photo

Facing page, upper left:
Bicolor fluorapatite on albite
1.2 cm wide
Telírio Mine, Linópolis District
Divino das Laranjeiras
Minas Gerais, Brazil
Luiz Menezes collection
Marcelo Lerner photo

Facing page, upper right:
Fluorapatite
Rio Grande do Norte, Brazil
Dilermando Rodriguez de Melo
collection, Guido Steger photo

Facing page, lower left:
Fluorapatite,4.3 cm wide
Urucum Mine, Galiléia, Minas
Gerais, Brazil
Álvaro Lúcio collection
Marcelo Lerner photo

Facing page, lower right:
Fluorapatite, 2.5 cm wide
Medina, Minas Gerais, Brazil
Luiz Menezes collection
Marcelo Lerner photo

Conselheiro Pena, Minas Gerais) also exploits a pegmatite that is embedded in granite. In 2006 a small cavity in this mine, better known for quartz and spessartine, produced a fine lot of flattened, pale purple fluorapatite crystals (some of which reached 5 centimeters) on albite.

Several pegmatites in the Linópolis district (Divino das Laranjeiras, Minas Gerais) produce outstanding brazilianite, eosphorite, and montebrasite crystals; and fluorapatite and hydroxylapatite crystals have been found in some of them.

The **Telírio Mine** is a source for green-yellow brazilianite crystals, light yellow to green-yellow montebrasite crystals, as well as tabular twinned colorless beryllonite crystals, all of which sometimes occur in association with tiny green fluorapatite crystals. And in 2004 the mine produced an interesting lot of bicolor (dark blue and colorless) fluorapatite crystals up to 1 centimeter in diameter on albite.

Some 5 kilometers northwest of Linópolis and on the western border of a large farm belonging to the Pomaroli family, the **João Firmino Mine** has produced large, pale yellow to green twinned montebrasite crystals. Several small bodies of secondary phosphates also occur within the pegmatite. These bodies contained intense blue fluorapatite crystals measuring up to 2 centimeters in diameter that are associated with a series of rare minerals species such as zanazziite, moraesite, beryllonite, ushkovite, atencioite, leucophosphite, eosphorite, and natrodufrenite.

Located about 1 kilometer outside the village of Linópolis, the **Lavra do Jaime** pegmatite produced several pockets bearing clusters of hydroxylherderite crystals, some as much as 8 centimeters long. Many of these pockets also contained tiny, prismatic, pale gray to medium blue hydroxylapatite crystals, with lengths up to 1 centimeter.

BORBOREMA PEGMATITE PROVINCE

The Borborema Pegmatite Province extends 200 kilometers from central Rio Grande do Norte to north-central Paraíba. These pegmatites have traditionally been worked for beryl and tantalite; tourmaline was not an important target until 1988, when the fabulous copper-rich, electric blue elbaite, commonly known as "Paraíba tourmaline" was found in **São José da Batalha**. The Equador and Pedra Bonita mines in the state of Rio Grande do Norte have hosted important finds of blue euclase and orange spessartine, respectively. And fluorapatite crystals are frequently found in many other pegmatites in this province.

The **Alto da Cabeça Mine** (Parelhas, Rio Grande do

Left: Fluorapatite and elbaite, 5.5 cm tall
Alto da Cabeça Mine, Parelhas, Rio Grande do Norte, Brazil
Bill Larson collection, Jeff Scovil photo

Above: Fluorapatite, 3.4 mm wide
Quintos de Baixo Mine, Parelhas, Rio Grande do Norte, Brazil
Reinhard Wegner collection, Marcelo Lerner photo

Facing page: Fluorapatite and elbaite, 11.4 cm tall
Alto da Cabeça Mine Parelhas, Rio Grande do Norte, Brazil
Steve Maslansky collection, Jeff Scovil photo

Norte), for example, has produced numerous fluorapatite pockets, including one containing zoned blue crystals and another with bipyramidal blue crystals, which measure up to 6 centimeters. And in late 2003 the mine yielded a cavity containing very fine dark blue fluorapatite crystals measuring up to 4 centimeters on albite or quartz. The apatite crystals are prismatic, with striated prism faces and terminations consisting of pyramidal faces with basal pinacoids.

The best specimen from this find consists of a 5.5-centimeter-long green elbaite crystal with a cluster of beautifully formed, blue fluorapatite crystals, up to approximately 2 centimeters long, perched on one end of it (photo above left). This as well as many of the other better specimens from the find at the Alto da Cabeça Mine were erroneously labeled as having come from Paraíba. Most of the fluorapatite crystals from this pocket were recovered as single crystals, but a few creative souls, of questionable ethical character, transformed the loose crystals into "matrix pieces" by gluing them onto quartz or albite crystal clusters, and covering the points of contact between the fluorapatite and "matrices" with tiny gray muscovite crystals. The same technique has been used to fake "matrix" euclase specimens for many years. Collectors are thus well advised to carefully examine specimens originating in northeastern Brazil.

The **Quintos de Baixo Mine** in Parelhas is a second important source for gem-grade cuprian elbaite. The fluorapatite from this mine is generally small but can be of excellent quality: bicolor or even triple-colored crystals have been recovered as have fine purple crystals. In the book *Minerals and Precious Stones of Brazil* (2010), these specimens were wrongly attributed to Equador, but the correct locality is Quintos de Baixo, which lies within the município of Parelhas.

The **Capoeira I Mine** (Parelhas, Rio Grande do Norte) is also a source for small blue fluorapatite crystals. The largest apatite-bearing pocket contained several hundred "miniatures." The most interesting find came in the early 2000s when a lot of hollow, pale pink fluorapatite crystals forming scorpion tail clusters measuring up to 3 centimeters were mined.

According to Professor Reinhard Wegner of the Universidade Federal de Campina Grande in Paraíba, the **Alto Feio Mine** (Pedra Lavrada, Paraíba) is Borborema province's main source of apatite. Several pockets in this mine produced hundreds of pale blue crystals, some of which were as much as 20 centimeters long. Wegner has a 30-carat faceted apatite that originated in this mine.

OTHER SOURCES

Other deposits in addition to those in the two major Brazilian granitic pegmatite provinces are important producers

of apatite specimens. Just 15 kilometers away from downtown Belo Horizonte and still within the city limits, the **Morro Velho** gold mine (Nova Lima, Minas Gerais) operated until 2003. The mine reaches a depth of 2400 meters and was for many years the deepest gold mine in the world. It worked a quartz-ankerite vein impregnated with gold-bearing iron sulfides and arsenides, and in the upper part of the mine it was not uncommon for miners to encounter cavities containing crystals of siderite, dolomite, ankerite, quartz, pyrrhotite, chalcopyrite, and pyrite as well as minor quantities of native gold, cubanite, ullmanite, pink fluorapatite, and red scheelite.

Morro Velho fluorapatite crystals are flattened, showing very short to almost indiscernible prism faces with dominant pyramidal and short pinacoid faces. The crystals are known to reach 5 centimeters in diameter. They are generally pale pink but the occasional specimen exhibits a more deeply saturated hue. The color of pink apatite in general, however, tends to be quite unstable: when exposed to sunlight, pink crystals fade to colorless in just a few minutes.

Ipirá in the state of Bahia has produced large blue prismatic fluorapatite crystals with pyramidal terminations but without visible basal pinacoids. Crystals up to 20 centimeters in length have been found in diopside-rich pyroxenites and syenitic pegmatites. These normally occur in massive white or orange calcite; thus, most of the apatite production came from the upper, weathered, parts of the deposit where the calcite had been leached away, and the fluorapatite crystals are found in residual clay or iron hydroxide. Crystals can be flawless and thus produce top quality, intense blue gemstones. But because the crystals are etched out of the encasing calcite, their faces are not as sharp as those of apatite that formed in open cavities.

In the early 1940s, the skarns of **Fazenda Firmeza** (São Tomé, Paraíba) produced some 300 tons of rough apatite crystals. The largest crystal weighs 60 kilograms and is in the collection of the Universidade Federal de Pernambuco. The miners that found the giant crystal were disappointed as they thought they had found beryl, but as it turned out they had expended considerable effort to recover "worthless" apatite.

Fluorapatite, approximately 3 cm long
Ipirá, Bahia, Brazil
Andrea Bartorelli collection, Marcelo Lerner photo

Apatite is also noted from four of Brazil's large industrial mines that exploit igneous alkaline deposits: the Jacupiranga Mine in São Paulo; the Araxá and Tapira mines in western Minas Gerais; and the Catalão Mine in Goiás. Unfortunately the apatite from all but the **Jacupiranga Mine** is fine-grained and with the exception of some colorless micro crystals found in small cavities at Jacupiranga, no collectible apatite has been recovered from these deposits.

CONCLUSION

Production of gem rough and mineral specimens from Brazilian pegmatites has declined steadily since the golden age of specimen production that spanned from the 1960s into the 1990s. This decline is due in part to difficulties involved with gaining mine and environmental permits, stricter labor and health regulations, and rising mining costs (especially as a result of the weakening of the U.S. dollar against the Brazilian real). The majority of Brazil's important pegmatites have yet to be exhausted, and new deposits will likely be discovered, so the future certainly holds promise for more outstanding finds of Brazilian apatite and other collectible minerals.

ACKNOWLEDGEMENTS: *The author wishes to extend his thanks to Carlos Cornejo, author of the comprehensive* Minerals and Precious Stones of Brazil *(2010), for generously sharing a number of the photos that appeared in that book.*

Apatite from the Andes

by Alfredo Petrov and Jaroslav Hyršl

Fluorapatite is one of the more beautiful minerals found in the **Siglo XX** tin mine, in the town of Llallagua (pronounced *Ya-ya-wa*). Colorless, purple and pink crystals of the species were at one time quite abundant. Some of the smaller crystals were even fairly gemmy. Crystals which almost appeared metallic—a result of being heavily included with fine jamesonite needles—were also found. The largest transparent to translucent colorless crystals known were collected by American mine engineer Mark Bandy in the Contacto vein in the 1940s. Those legendary crystals reached 5 centimeters and exhibited nine crystal forms (Bandy 1976)!

Bandy favored his apatite crystals over any of the other minerals he collected during his several years working in Llallagua. He was delighted when miners broke into an apatite pocket that was big enough to sit in! Bandy spent many hours carefully recovering apatite specimens from his big vug, some would say at the expense of his company responsibilities. And after Bandy's third day collecting the pocket and with the vug only half worked out, one of his coworkers, upset about his spending company time collecting minerals, threw a stick of dynamite into the pocket and blew it apart. Bandy was heartbroken; still many of the finest Bolivian apatites in collections today came from this pocket.

Llallagua apatite exhibits compositional zoning that leads to unusual and beautiful luminescence and optical properties. Most are strongly luminescent with pink, orange, and blue-white zones. Some exhibit a strong face specific fluorescence: orange on prism {100} faces and bright violet on pinacoid {001} faces (Rakovan 2002). They formed 43.8 million years ago, give or take 4.7 million years (Rakovan et al. 1997). Like apatite from the Ashio Mine in Japan (see page 98), many of the crystals from Llallagua are not truly hexagonal but are optically complex (and beautiful) under the petrographic microscope (Rakovan and Reeder 1994).

The most recent find, in 2002, yielded tabular apatite crystals to 3 centimeters wide but only 4 millimeters thick. The crystals are completely covered with yellow wavellite crusts. They are light pink in daylight or tungsten light but are lemon yellow under fluorescent tubes, a phenomenon which indicates a high content of certain rare earth elements (Rakovan and Reeder 1996). Fluorapatite is one of the oldest minerals in the tin veins, and by leaching and decomposition, it gave rise to the numerous younger phosphates, especially wavellite, vauxite, and paravauxite that are so well known from Llallagua. Pale orange perimorphs of wavellite after apatite have sometimes been found as have apatite crystals that have been partly replaced by gypsum or vivianite.

The **Cerro Grande** tin mine in Cochabamba Department is a lesser known locality for fine apatite specimens. Colorless apatite crystals up to 3 centimeters long display a beautiful blue to peach-colored fluorescence. Cerro Grande specimens are easily recognizable because they sit on crusts of tiny drusy cassiterite, quartz, and siderite crystals. The apatites are invariably by far the largest crystals on the specimens.

Agustín Arando and Alfredo Petrov prospected for apatite crystals on **Cerro Huañaquino**, a few kilometers north of the city of Potosí, in 1995. Agustín staked a claim over an apatite vein for specimen mining under the name "**David Mosiah**." The pair discovered single or complex parallel growths of prismatic apatite crystals up to about 10 centimeters long composed of a carbonate-bearing outer opaque creamy white zone with a silky surface luster. The crystals fluoresce weakly whitish under both shortwave and longwave ultraviolet light, but their cores fluoresce orange in shortwave ultraviolet light, occasionally showing iridescent reflections from their basal cleavages. The cores of these crystals are transparent to translucent, light gray to light yellow, with subvitreous luster. Some crystals have holes through their centers, parallel to the *c*-axes. Such is the case with the largest crystal ever found at this locality. The specimen is currently in the collection of Rock Currier, who has put a rope through the central hole so that it can be worn as a big clunker of a pendant!

Fluorapatite, 2.7 cm wide
Siglo XX, Llallagua, Potosí Department, Bolivia
Les Wagner collection, Jeff Scovil photo

As is the case with those from Cerro Grande, apatites from Cerro Huañaquino are by far the largest crystals in the deposit and are associated with much tinier crystals of feldspar, rutile, magnetite, barite, etc. The black spots commonly seen peppering the apatite crystals are reticulated rutile pseudomorphing platy ilmenite crystals (Petrov 1997). Smaller, thinner prismatic apatite crystals are sometimes found in association with the fine magnetite specimens abundant at other claims on this mountain. The specimens generally came out from about 1999 to 2007.

Siglo XX, Cerro Grande, and Cerro Huañaquino are the only significant apatite specimen producers in Bolivia, although a few other well known tin and silver mines have yielded lesser crystals. Small, well-developed crystals show up above the city of Potosí at **Cerro Rico**, where tin-silver veins occasionally yield apatite. One such find in the 1950s in the Krause vein netted nice small crystals associated with some of Bolivia's very best phosphophyllite crystals. Lustrous, white crystals to one centimeter and accompanying pyrargyrite crystals were recovered from the Potosí vein in 2002. Well-formed simple prismatic crystals to about one centimeter long occur rarely with siderite, quartz and cassiterite at the **Viloco** tin mine in La Paz Department. And small pale violet crystals just a few millimeters in size sometimes show up in sulfosalt vugs at the **San José** and **Poopó** mines in Oruro Department. Flat white crystals to 3 centimeters are known from the **Animas Mine** in Potosí Department.

One last and quite extraordinary Bolivian apatite, or perhaps we should say "missing apatite," merits mention: the hollow black perimorphs of cassiterite after tabular or columnar apatite crystals from the **Chorolque Mine**, one of the world's highest mines, where tin, tungsten and gold are mined at elevations up to 5,486 meters!

Although Peru is primarily known for ore minerals, the country is also home to several interesting apatite localities, each of a different genetic type (Hyršl et al 2011). By far the best known and the most productive is the large **Huanzala** skarn deposit, the world's most productive source of pyrite crystals. Apatite from Huanzala is usually small, forming tabular or prismatic colorless, white, or pink crystals generally up to one centimeter long and only rarely reaching about 2.5 centimeters. They are popular among collectors for their strong, sometimes zoned, white fluorescence visible under ultraviolet light and for their unique association— they usually form in nice pyrite vugs.

The **Turmalina Mine** near Piura in far northern Peru is a pneumatolytic, breccia-pipe copper-molybdenum deposit. The locality is virtually unknown internationally, but in the 1990s it yielded fantastic scheelite crystals up to 18 centimeters wide as well as Peru's best apatite specimens, consisting of white to gray columnar apatite crystals to 10 centimeters associated with crystals of arsenopyrite, molybdenite, chalcopyrite, quartz and schorl.

Another of Peru's important apatite occurrences is in the polymetallic ore district **Julcani**. The district produced near-colorless to pale yellow prismatic apatite crystals up to 2.5

centimeters that formed on sulfides, mainly arsenopyrite. The most recent important find came in 2008 and yielded nice short columnar apatite crystals to 2 centimeters growing together with well-crystallized chalcopyrite and sharp siderite rhombohedra. Julcani apatite often displays a nice zoned fluorescence under ultraviolet light.

Finally, small white apatite crystals are known from the **Pasto Bueno** tungsten deposit (also a source of splendid hübnerite crystals and rhodochrosite specimens) and from a skarn deposit at **Pampa Blanca**, which is also the source of thousands of Japan-law twinned quartz and of epidote. Apatite from both of these deposits is quite rare.

Minor apatite has been noted throughout the emerald deposits of Colombia but there is one particularly noteworthy source, the **La Marina** emerald mine, in the municipality of Pauna, Boyacá Department. A black shale hosted hydrothermal vein deposit, La Marina Mine is geologically similar to its more famous neighbors Muzo and Chivor, but La Marina is set apart by the beautiful pink fluorapatite and exceptional euclase crystals that it produces. Both of these may be found in association with albite, calcite, and emerald. Hot pink fluorapatite on rich green beryl ("emerald") make for particularly striking specimens.

The apatite is typically of a simple tabular habit, although equant to slightly elongate crystals have been found, and are dominated by the first order prism and pinacoid forms. The intensity of the pink color varies, mostly with the size of the crystals, and larger specimens can appear quite saturated. Although usually around a centimeter in size, individual crystals as large five centimeters have been found. They commonly form aggregates of stacked, parallel crystals.

Upper:
Fluorapatite, 2.6 cm tall
La Marina Mine, Pauna,
Boyacá Department, Colombia
Well Arranged Molecules specimen
Jeff Scovil photo

Right:
Fluorapatite with arsenopyrite, 4.5 cm wide
Julcani District, Huancavelica
Peru
John Rakovan collection
Jeff Scovil photo

Facing page:
Fluorapatite, 14.9 cm wide
Cerro Grande Mine,
Cochabamba Department,
Bolivia
Marty Zinn collection
Jeff Scovil photo

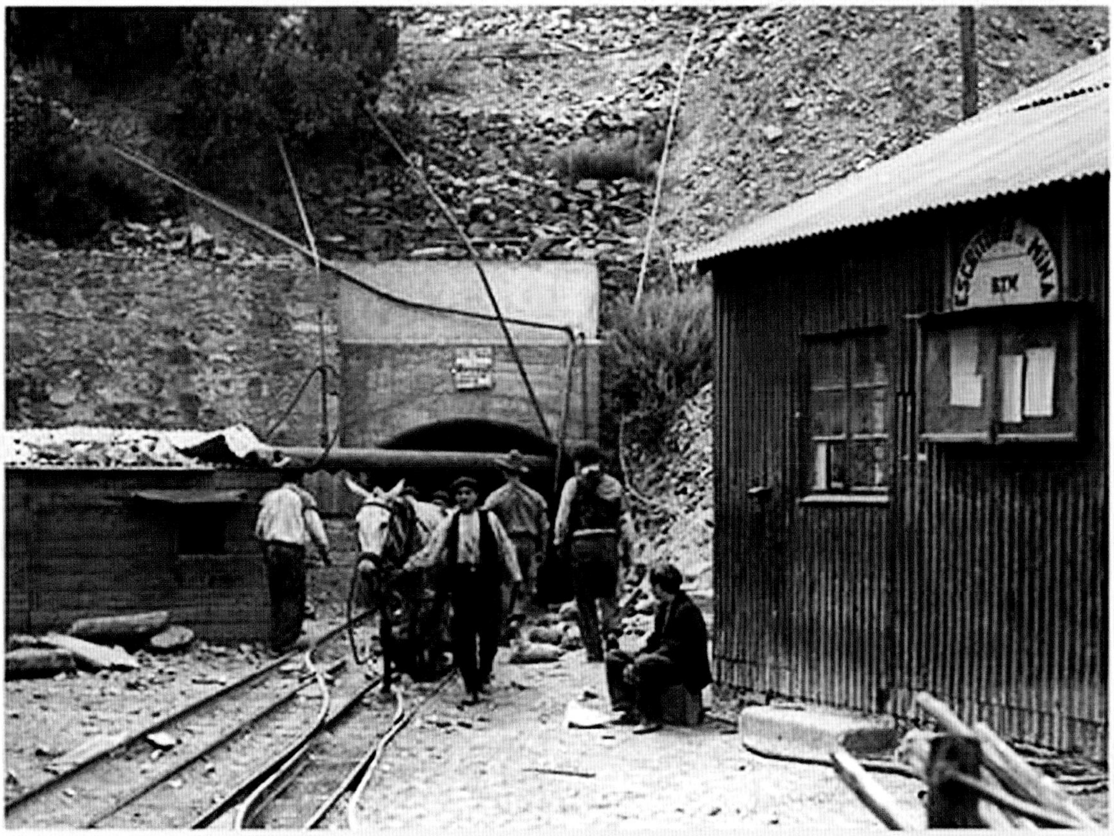

The Panasqueira Mines: One Hundred Years of Fluorapatite

Carles Curto Milà and Jordi Fabre i Fornaguera

From the moment they made an appearance on the international stage, apatite specimens from Panasqueira became mineralogical classics. For decades, literally thousands of good to exceptional specimens have flowed into great (and not so great) collections around the world. The magnificent crystallization, the size and morphology of the crystals, the variety of color, and the diverse and interesting associations have and continue to compel collectors to acquire Panasqueira apatite specimens, which have become benchmarks in any collection of renown.

Management changes at the mining company and newly imposed regulations have restricted the extraction and sale of specimens by the miners; thus, production of collectible examples of fluorapatite and other Panasqueira species seems to have declined in recent years. Still, growing demand for tin and tungsten have stimulated the company to work new exposures near Vale da Ermida and other areas of the mine complex. Given that and the ubiquity of the species, prospects for new finds of fluorapatite are promising.

REGIONAL GEOLOGY

Panasqueira is one of the world's largest and best known tungsten deposits. Located around the tiny town of Aldeia de São Francisco de Assis, which falls within the municipality of Covilhã, Castelo Branco district in central-western Portugal, about 300 kilometers northeast of Lisbon, the deposit falls within the "Tin Belt," one of the most important metallogenic and mining regions of Europe. The Tin Belt is marked by a series of granitic massifs, which have domes of more highly evolved and volatile-rich intrusions protruding from their roofs. These domes are cut and surrounded by metal-bearing

Above: Vintage photo of an entrance into the Barroca Grande Mine, part of the Panasqueira mine complex

Facing page: Fluorapatite, 13.5 cm tall
Panasqueira mine complex, Aldeia de São Francisco de Assis
Municipality of Covilhã, Castelo Branco district, Portugal
Marcus Budil specimen, Jeff Scovil photo

Top: Fluorapatite with arsenopyrite, 4.8 cm wide
Panasqueira mine complex, Aldeia de São Francisco de Assis
Municipality of Covilhã, Castelo Branco district, Portugal
C & C Forrester collection, Jeff Scovil photo

Middle: A fold exposed on the surface in rocks in the Tin Belt, just outside of the Panasqueira mining concession
Josè Arce Diaz photo, June 2012

Bottom: Quartz vein in the Panasqueira mine; note the muscovite and schorl-dravite along the contact between the vein and the wallrock, Josè Arce Diaz photo, June 2012

stockworks, veins, and greisens. Panasquiera's geology and metallogeny have been exhaustively studied and described (Ascensao Guedes 2002; Azevedo and Sáinz de Baranda, 1997; Bloot and De Wolf 1953; Bussink 1984; Cardoso d'Orey 1971; Conde, et alt. 1971; Cotelo 1944; Foxford, et al. 1991; Kelly and Rye 1979; Reis 1971; Thadeu, 1951); thus, we will limit ourselves to a general discussion of the regional geology as a context for our principal focus—those aspects directly related to the distribution and genesis of fluorapatite.

The rocks of the area range in age from Upper Precambrian to Cambrian and were successively affected by the Caledonian, Hercynian, and Alpine orogenies. The sequence has been cut by granitic batholiths that caused regional low-grade thermal metamorphism, which is locally overprinted by contact metasomatism. The rocks locally retain clear evidence of their marine sedimentary origin as shales with lesser interbedded greywacke and quartzite. Where affected by regional metamorphism, their origin is obscured by their conversion to schistose slate, with biotite and chlorite nodules as well as rarer andalusite (variety chiastolite) and cordierite.

The granites are affected by greisen alteration (which is poor in sulfides) and a cap of hydrothermal silica, which locally reaches 14 meters in thickness. The granite batholiths are cut by north-south-trending, subvertical mafic (dolerite) dikes and younger aplites that are genetically related to the granites. These intrusions are in turn cut by lenticular barren quartz veins called *seixo bravo* (literally "quartz rough"), which are concordant with the foliation of the slates.

MINING GEOLOGY

The dense stockwork of quartz veining that hosts the mineralization controls to a significant degree the deposit's mineralogy. The veins show subhorizontal structures principally mineralized by ferberite and cassiterite, with particular enrichment in sulfides. These structures may be crossed by later, steeper structures that are also mineralized. The mineralization was deposited on the walls of the veins in successive parallel layers that show a clear depositional progression of different stages of filling from the walls into the center of the veins. The layer along the wallrock contact consists of a mixture of muscovite and tourmaline (schorl-dravite), topaz, and cassiterite. Ferberite was deposited atop this in parallel growth, which occupies most of the center of the vein. Carbonates and minor late-stage sulfides mark the final phase of infilling.

This style of vein filling created abundant open drusy or crystal-lined voids that provided the surfaces and space needed for large crystals to grow. In the drusy pockets the principal minerals are ferberite, fluorapatite, cassiterite, and arsenopyrite, which generally grown on tiny acicular schorl-dravite crystals or leafy muscovite. This assemblage can be covered by scattered or concentrated sulfide crystals including marcasite, pyrite, or chalcopyrite; and carbonates including siderite, calcite, and dolomite. The mineralization of the crystal-lined voids is slightly different in specific areas of the deposit, where it is characterized by well-developed quartz crystals with associated fluorapatite, siderite, arsenopyrite, and pyrite-marcasite.

Thadeu (1951) divided the deposit into six areas based on the morphology of the veins, structural controls, and

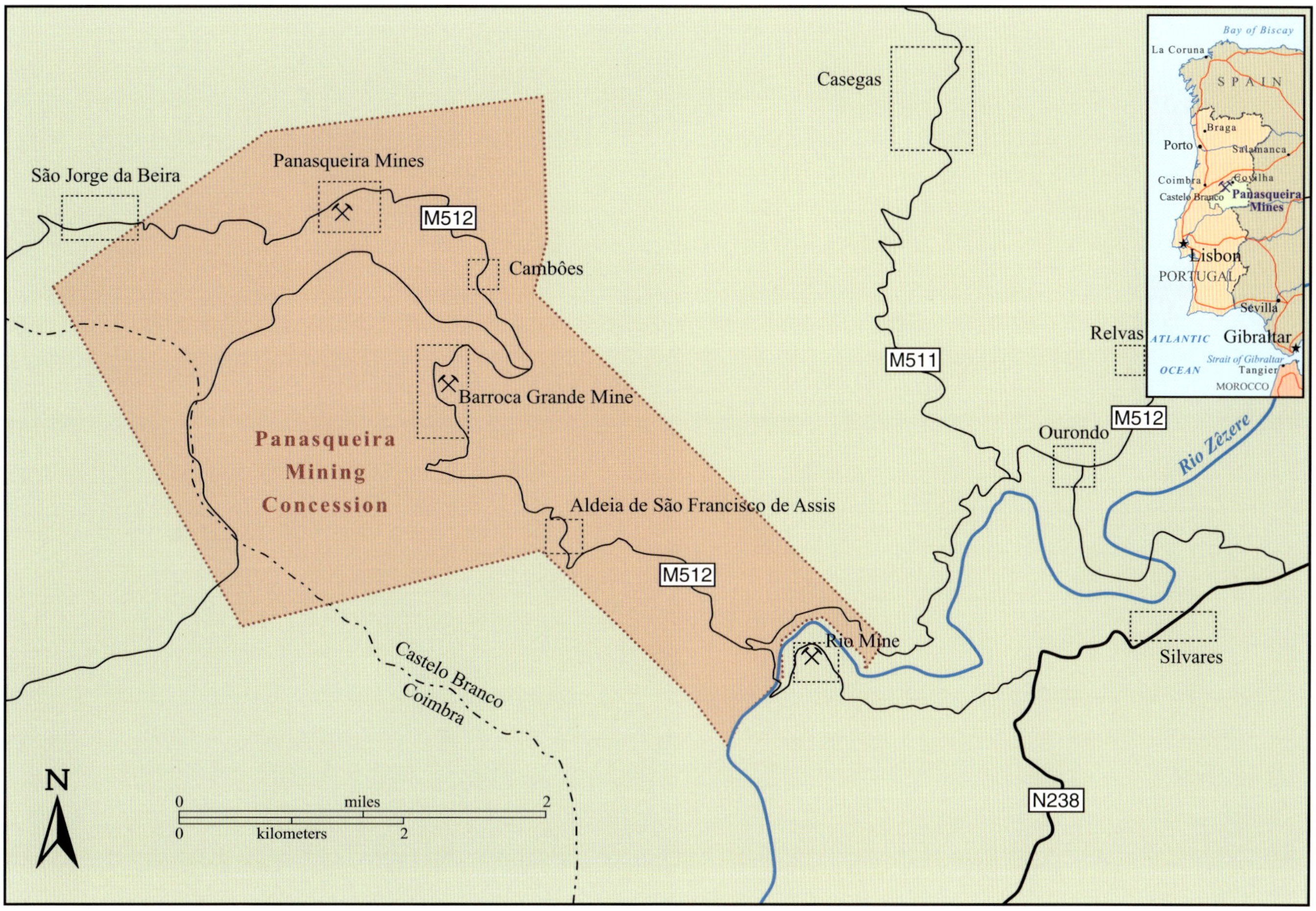

Map: The Panasqueira Mining Concession
William W. Besse cartography

mineralogy. The areas that have been worked most extensively at various times are: **Panasqueira-Barroca Grande**, **Gorga Seca-Alvoroso-Veia Branca-Giestal**, **Lomba da Cevada**, **Rebordões-Seladinho**, **Fonte das Lameiras**, **Vale das Freiras-Vale da Ermida**, and **Cabeço do Pião**. The majority of the fluorapatite crystals came from the Vale das Freiras-Vale da Ermida areas, especially the Panasqueira (*sensu strictu*) and Barroca Grande areas. In these areas fluorapatite is ubiquitous, although irregularly distributed, and appears more abundant in Panasqueira and Vale da Ermida, although it is much scarcer in the deep levels of Barroca Grande (Azevedo & Calvo 1997).

In general, most fluorapatite deposition occurred during the initial high-temperature, high-fluorine stages of deposit formation with subsequent deposition during primary muscovite crystallization and even later ferberite formation (Knutson, et al. 1985). The principal fluorapatite concentrations show a clear alternation between periods of extensive deposition and periods of almost no growth. There are also many examples of later fluorapatite remobilization and redeposition, resulting in the wide variety of mineral associations and temporal relationships seen between fluorapatite and other species in the Panasqueira deposit.

A BRIEF HISTORY

Ancient ruins present clear evidence that the Panasqueira deposit was mined in Roman times and later when present-day Portugal was under Muslim rule. Tin appears to have been the only target at that time. The old workings were small and were limited to superficial deposits. The first written reference to mining at Panasqueira was in the 1889 catalog *Minas da Secção Exposição da Lisboa*, which summarized mining activities for the years prior to its publication (Vaz Leal 1942; Azevedo 1997), but these activities were almost certainly insignificant.

Sociedade de Minas de Volfram em Portugal Lisboa por a firma Almeida, Silva Pinto & Comandita, Exploradora de Minas de Volfram existentes no Fundão e Covilhã, com sede en Lisboa was the first official mining concession to have registered with the Diario do Governo (in 1898) for the purpose of extracting tungsten from Panasqueira as well as from Cabeço do Pião.

In 1901 Count Burnay acquired the concession and, retaining the brand Silva Almeida Pinto & Co., leased the mine to a British company, which operated the mine for less than one year. Burnay worked the deposit until 1911, when it ceded the property to Wolfram Mining & Smelting Co., Ltd. The mine ran under that company until 1928 and then Beralt Tin & Wolfram Ltd. took over, exploiting the ores solo until 1973. In that year, the Portuguese government became invested in the operation, which was renamed Tin & Wolfram Beralt Portugal, SA; the majority stake-holder in the new company was the Canadian firm Primary Metals. Between 2007 and 2008 the Japanese company Sojitz launched a successful takeover bid for Primary Metals. Sojitz took control of the Panasqueira operation and changed its name to Tin & Wolfram Beralt Sojitz (Portugal) SA.

THE FLUORAPATITE

With a relatively complex, species-rich mineralogy, Panasqueira is famous for its mineral specimens, some of which are among the best of their species in the world. The top-of-the-line crystallization includes fluorapatite, ferberite, cassiterite, arsenopyrite, and siderite. Other species such as quartz, marcasite, chalcopyrite, pyrite, sphalerite, calcite, dolomite, and fluorite—64 in all—are present to a greater or lesser degree. And Panasqueira is the type locality for two of these: the rare species panasqueiraite (Isaacs and Peacor 1981) and thadeuite (Isaacs et al. 1979).

The genesis, morphology, chemistry, and color of Panasqueira fluorapatite have been widely studied and described, first by Bloot and De Wolf (1953), with numerous subsequent studies referenced in professional and popular journals in the years hence. The presence of other members of the solid solution series, namely chlorapatite and hydroxylapatite, has been the source of speculation, as characteristics such as the fibrous appearance and/or yellow to almost white color of some crystals have been thought to be diagnostic of chlorapatite or hydroxylapatite, but analyses have repeatedly shown every sample to be fluorapatite (Dunn 1977; Hochleitner and Weiss 2005; Kelly and Rye 1979). Similar misconceptions about visually diagnostic characteristics of chlorapatite and hydroxylapatite have been proposed for specimens elsewhere: Chichibu, Japan, for example (John Rakovan personal communication).

Panasqueira fluorapatite crystals exceeding 10 centimeters are not particularly rare, but their longest dimensions more commonly range from 1 and 3 centimeters. Their habits vary, but crystals predominantly take flattened tabular forms. Prisms are also relatively common, with the ratio between their lengths and widths in the 2 to 1 range. The rare crystal may even be needle-like or fibrous.

In general the crystals are morphologically simple. The first order prism {100} and pinacoid {001} are clearly more developed than are other faces. Crystals are commonly modified by the second order prism {110}, the development of which occasionally equals that of the first order prism. Less common than the pinacoid {001} and the prism {100} are the dipyramid forms that modify crystals of simple habit and in rare instances are dominant forms. The steeper dipyramids {101} and {102} are more common, although some crystals exhibit more shallow {103} and {104} dipyramid faces. Of course, nothing is ever totally simple: close inspection of some crystals can reveal rich and complex crystallographies that include more uncommon dipyramids (positive and negative) with very different inclinations (Azevedo and Calvo 1997; Lagerwey 1977).

Of note are frequent doubly terminated crystals, often with a pseudo-hemimorphic appearance due to the development of different faces on the two ends of *c*-axis. In other instances the crystal morphologies may form partnerships of parallel intergrowths making complex polycrystalline terminations.

Individual crystals as well as aggregates of numerous individuals are found. The groups are usually irregular, sometimes forming "nests" of crystals that are coated on their undersides and are bordered in part by calcite or siderite. Aggregates consisting of parallel scepters and inverse scepters as well as curious sheaf-like or "mushroom" crystal aggregates have been observed (Ascensao Guedes 2002).

Habits of Panasqueira Fluorapatite
Mines near Silvares, Castelo Branco, Portugal

Above: Aggregate of short prismatic fluorapatite crystals 5.8 cm tall, Marty Zinn collection, Jeff Scovil photo

Upper right: Long prismatic fluorapatite crystal, 8.5 cm tall Arkenstone specimen, Joe Budd photo

Lower right: Stacked tabular fluorapatite crystals 6.5 cm long; Frederick Escaut specimen, Jeff Scovil photo

Below: Doubly terminated, pseudo-hemimorphic, and inverse sceptered fluorapatite, 3.6 cm long Private collection, Deusedes & Fabre photo

Facing page: Parallel tabular fluorapatite crystal group with siderite, 3.9 cm; Carles Millán collection, Jeff Scovil photo

COLOR AND COLOR ZONING

A review of prominent collections, mineralogical publications, and a simple internet search reveals the wide range of morphologies and colors seen in fluorapatite from Panasqueira. The dominant color, which is clearly independent of habit and morphology, is green, with shades varying from bottle-green to olive to gray-green. It has been speculated that green in Panasqueira fluorapatite is brought about by the presence of ferric iron (Fe^{2+}), and that the iron-imparted color is made progressively paler by the bleaching action of Nd^{3+} and/or Ce^{3+} (Knutson et al. 1985). Some but not all of the green apatite from the site fluoresces mustard-yellow under both long and shortwave ultraviolet light. This fluorescence is more pronounced in color-zoned crystals and is activated by manganese (Mn^{2+}) (see page 10).

The violet palette of colors is the second most common seen in fluorapatite from the deposit. Crystals range from blue-violet to lilac or violet-blue; their color is generally pale but, particularly in the case of the violet, may be deeply saturated (Ascensao Guedes 2002; Azevedo and Calvo 1997; Bancroft 1984; Gaines, et al. 1971). Lagerwey (1977) suggests that rose, red-purple and blue-purple colors are the result of Mn^{3+} ions in the calcium sites of the apatite structure. Blues and blue-greens appear to be the result of color centers. Colors in the violet spectrum, unlike those in the green, are usually concentrated at the intersections of the pinacoids {001} and bipyramids {110} such that the outer edges of thin, tabular, violet crystals tend to be more deeply saturated than their cores.

Other colors of fluorapatite found at Panasqueira include pale shades of blue and pink, both of which are rare, as well as white and yellow. Colorless crystals are relatively abundant

Right: Fluorapatite with arsenopyrite, 6 cm wide
Brian Kosnar collection
Joe Budd photo

Below right: Fluorapatite and quartz on siderite and arsenopyrite, 6.7 cm long
Marty Zinn collection
Jeff Scovil photo

Facing page, upper left: Fluorapatite, 4.8 cm tall
Jordi Fabre collection
Jeff Scovil photo

Facing page, upper right: Fluorapatite, 10.2 cm tall
M. Carré collection
Jeff Scovil photo

Facing page, middle: Fluorapatite on quartz; 3.3 cm wide
Jordi Fabre collection
Jeff Scovil photo

Facing page, lower: Color-zoned, 3 cm wide fluorapatite group on muscovite
Jordi Fabre collection
Jeff Scovil photo

but are generally small. And a few crystals exhibit different colors (fine green tones to violet) under different light sources (e.g. incandescent, fluorescent, or ultraviolet).

Color zoning is also one of the most striking and attractive characteristics of Panasqueira fluorapatite. Tabular to short prismatic fluorapatite from this locality commonly exhibits concentric color bands, while prismatic crystals tend to exhibit "phantoms" of color. At times the color toward the centers of crystals, particularly the violets and pinks, is more intense than toward the peripheries.

Although some authors (Lagerwey 1977, Knutson, et al. 1985) have hypothesized that certain types of color-zoning are associated with specific components of a crystal, such as at a contact between the pinacoid and dipyramid; in reality, apatite color zones cannot be conclusively attributed to morphology or any other characteristics of the crystals. Crystals with alternating opaque, matte, pale colored faces; bright, white faces; and dark green faces are sometimes found. The colors of these faces are undoubtedly related to the crystal morphology in some way.

Concentric zones around the *c*-axis (in other words concentric zones within the prism {100} and {110} sectors) are the color zones most frequently seen in Panasqueira fluorapatite and are particularly apparent in tabular and short prismatic examples. The crystals that exhibit these zones, locally known as *olho de boi* ("porthole"), almost always show two or at most three dominant bands (photos above and facing page, lower). The most common configuration consists of prism faces with dark green cores and thin, pale green to sometimes colorless rinds that extended to the pinacoid faces. The most striking crystals show bands with distinctly different colors, such as a dark central area, a transparent green intermediate band and a colorless rind.

Left: Fluorapatite with arsenopyrite and siderite
5.2 cm wide
Joan Massagué collection
Jeff Scovil photo

Below left: Fluorapatite with fluorite, 4.2 cm tall
Steve Smale collection
Jeff Scovil photo

Below right: Fluorapatite on ferberite
5.2 cm wide
Mike Keim collection
Jeff Scovil photo

Facing page: Fluorapatite with quartz, pyrite and muscovite
6.1 cm wide
Joan Massagué collection
Jeff Scovil photo

ASSOCIATIONS

No less interesting from the standpoint of both the collector and the researcher are the mineral species with which apatite is found, as these make specimens particularly attractive and mineralogically interesting. While fluorapatite is more abundant in certain areas of the site than in others, it is ubiquitous and mobile and has crystallized before, after, or simultaneously with other species. At Panasqueira fluorapatite can occur in combination with numerous species including ferberite, arsenopyrite, cassiterite, siderite, and to a lesser extent quartz, calcite, and sulfides such as sphalerite, chalcopyrite, pyrite, and marcasite (Ascensao Guedes 2002; Azevedo and Calvo 1997; Cardoso d'Orey 1971).

Sizeable crystals of quartz and/or ferberite are commonly associated with or serve as matrix for later-stage fluorapatite, which may also be found on matrix carpeted with cassiterite, muscovite, and/or acicular schorl-dravite. Druses of late-stage carbonates and sulfides, especially calcite, pyrite, marcasite, and less commonly, dolomite and sulfides such as löllingite, stannite, and chalcocite frequently coat the entire assemblage of minerals at Panasqueira. Tiny crystals of pyrite that form on the terminal faces of fluorapatite make particularly attractive specimens. And in rare cases, heavily corroded fluorapatite is found intergrown with muscovite rosettes. Associations with marcasite are relatively uncommon in most areas of mineralization, although they have been observed in many samples from various areas of Vale da Ermida and Barroca Grande. Specimens from these areas are often partly coated with marcasite, and small fluorapatite crystals are dispersed on the faces of the crystals.

Apart from the typical associations, fluorapatite may also be found with many of the other species known from the mine including chalcocite (Vale da Ermida), chalcopyrite (sometimes with sphalerite), and topaz (Vale da Ermida, Barroca Grande). Fluorapatite seems to form at the expense of fluorite at the mine, making the latter species a rarity at Panasqueira. When found, the fluorite is usually in cavities and is sporadically associated with quartz, siderite, chalcopyrite, pyrite, and fluorapatite.

Fluorapatite and quartz are commonly included with other minerals, particularly ferberite. Such inclusions may be fully embedded in the host crystal or may be only partly included. Nearly centimeter-long ferberite inclusions in fluorapatite are known. Fine, acicular crystals of schorl-dravite series tourmaline can impart a bronzed appearance to the apatite. Small, silvery tassels of muscovite are clearly visible in many crystals, as are small, deformed crystals of arsenopyrite. Small, bright, cassiterite twins can be completely or partly included in fluorapatite, but the species is more commonly seen having formed on the faces of apatite crystals than inside of them. On the other hand, fluorapatite is not commonly included in other species at Panasqueira, although from time to time some, fluorapatite is found as an inclusion in transparent quartz crystals which when included are often also partly dissolved.

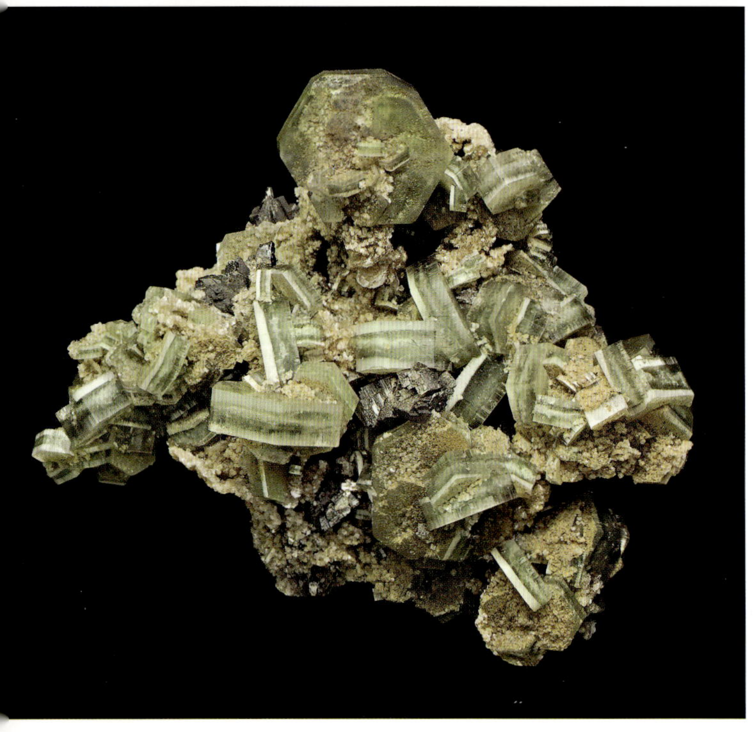

Upper: The ubiquitous carbonate coating has been partly removed (probably by air abrasion) from this 8.5 cm long cluster of purple fluorapatite crystals.
Houston Museum of Nature and Science collection
Jeff Scovil photo

Lower: Late-stage carbonates aesthetically encrust a 5.8 cm wide aggregate of tabular, pale green fluorapatite crystals on arsenopyrite. Jordi Fabre collection and photo

COATINGS AND DAMAGE

A certain percentage of specimens (from any locality) on the collectibles market are subject to repair, manipulation, or even forgery, and those from Panasqueira, specifically specimens of its famous fluorapatite, are no exception.

When it is first recovered, Panasqueira fluorapatite is commonly coated, at least in part, with mica, sulfides, or carbonates, the latter being, as noted above, especially common. The parts of the crystals that are exposed can appear quite promising and the temptation to remove the coating and expose the entire crystal is compelling. Keep in mind that one or more of the mineral species that formed the coating may have as it formed reacted with the fluorapatite such that the specimen may be damaged in the course of uncovering what may turn out to be a corroded surface.

It is generally inadvisable to remove coatings by mechanical means as these invariably cause damage to the crystals beneath. But because fluorapatite can be sensitive to acids that might otherwise easily dissolve the encrusting carbonates, miners nevertheless try to chip away the coatings with a *faca* ("sharp knife"). Sadly, use of this "advanced technology" has caused untold damage to specimens.

For years, damage as well as exposed, corroded surfaces had been more or less tolerated—written off as a necessary cost of the difficulties associated with specimen recovery. And damaged crystals can be readily seen in any collection built before the 1980s. In fact, it is unusual to find a specimen that does not have some (or many) damaged edges or dull surfaces.

Of the 15,000 specimens in the well-known Folch Collection, for example, more than 1,000 are from Panasqueira. Although the collection was largely amassed between the 1960s and 1980s—a time when the size of a specimen mattered more than its condition—Mr. Folch sought smaller, perfect examples of specimens whenever he could. Nonetheless, the vast majority of the Folch Panasqueira fluorapatite specimens show the inevitable dings or breaks.

Today, the high value placed on crystal perfection coupled with improvements in cleaning technologies have incentivized miners to more carefully recover and preserve specimens that they encounter *in situ*. On the other hand, modern mining methods are more mechanized, such that miners are less likely than before to encounter specimen-bearing pockets that have not been exposed to blasting or excavating equipment, both of which have the potential to do far more damage to specimens than a *faca*.

FAKES AND REPAIRS

The demand for and high dollar value of perfect specimens and the difficulties associated with obtaining them has stimulated development of an "underworld" of specimen repair at the mine. This industry has always existed in some form, but it has grown significantly in recent years, such that new finds from the mine need to be carefully examined for signs of repair, restoration, or out and out counterfeits.

The practice of faking specimens is not necessarily more sophisticated at Panasqueira than at other localities, but the nature of the deposit has a tendency to complicate the task of recognizing the ruse. At this locality, tectonic activity causes large numbers of crystals to break naturally. Because of the

Above: Mine tailings are a colorful backdrop to the little town of Aldeia de São Francisco de Assis in Castelo Branco, Portugal Tina Alves photo

Right: A sign marks the way into "Barroca Grande, Land of Wolfram (tungsten)"; Jordi Fabre photo

mobility of the fluids at the deposit, these broken crystals tend to recrystallize ("heal"), albeit rudely; thus it can be difficult to distinguish between natural breaks and "repairs" and those made by the human hand. Furthermore, naturally broken and rehealed crystals are also subject to being encrusted by later-stage carbonates, which tend to cover, cement, or weld everything in their paths. Noting this gross, natural cover-up, some of the more gifted forgers have taken to concealing damage caused during extraction or subsequent handling with manufactured "coatings" of carbonates or sulfides. Given that the undersides of the coatings often have smooth surfaces, having formed on top of smooth crystal faces, pieces of coating fit nicely over repairs and other manipulations. And as muscovite is ubiquitous at Panasqueira, tiny flakes are likewise used to conceal gaps of the products of the finest of craftsmen.

There are obviously great variations in the quality of the various forgers' work—anything from fluorapatite crystals glued rudely to quartz or ferberite to fine pieces created by skilled artisans is available. But in general an attentive eye, a good loupe and some healthy skepticism are enough to recognize most fakes. A thorough study of the specimen in question under a good light (preferably daylight) and with a little magnification are sufficient to discern bubbles, hairs, glue, or crystals in rare or absurd positions, any of which are indicative of a forgery. When in doubt, one can always immerse the specimen in an organic solvent, acetone for example, as thus far, any glues used to create Panasqueira fakes are sensitive to this type of solvent.

The intention of this section is in no way to alarm our readers nor to make them feel insecure about the veracity of the magnificent specimens from Panasqueira, as the vast majority are completely natural. Only a fraction of the total specimens that come to market are manufactured, and many of those that would be sold remain in Panasqueira because knowledgeable collectors and dealers reject them and rebuke those who try to sell them. However, as specimen prices rise and the practice appears on the upswing the world over, let this serve as a reminder that a critical eye and a closer look are always a good idea.

ACKNOWLEDGMENTS: *The authors wish to extend their thanks to Dr. Joan Viñals Olià, for his insights into sulfide relationships and to Inma Ramos Márquez, whose research improved this text.*

The large icy mass of the Rhônegletscher is the river Rhône's primary source of water. Although quite impressive, the ice has been melting at an alarming rate and is predicted to disappear by the end of this century. Apatite specimen-bearing Alpine clefts are found in the area of the receding glacier.
Mark van Laere photo, October 2009

Superb apatite crystals have been found in many localities scattered throughout the European Alps. The classic apatite-producing sites in Europe are mainly in Alpine clefts, especially those in Switzerland and Austria. Choice specimens have been collected since the early nineteenth century by both institutions and private collectors. Fortunately, a large number have been preserved, studied, and placed on display, such that excellent examples can be seen in nearly every major museum worldwide.

SWITZERLAND

Alpine clefts host the most significant Swiss mineral formations. These clefts are generally concentrated in the central Swiss Alps and can be up to 5 by 2 by 1 meters; only very rarely have larger clefts been found. Typically the clefts tend to be elliptically flattened perpendicular to the schist orientation of the surrounding rock. Quartz, orthoclase (variety adularia), albite, chlorite, and calcite are the most ubiquitous minerals in the clefts, whose parent rocks are granite, gneiss, granodiorite, syenite, amphibolites, and various schists.

Two of the most productive mineralized Alpine cleft regions in Switzerland are the Aar and Saint Gotthard massifs. One of the most widely disseminated minerals throughout these massifs is apatite. It plays an important role in the classification of cleft paragenesis. Work done by Gustav Rose in 1827 on apatite specimens from Saint Gotthard showed them to be specifically fluorapatite.

Apatite crystals of the clefts are usually clear and rich in forms ranging from tabular to long prismatic. In terms of size, the tabular forms are microscopic to about 6 centimeters wide, and prismatic forms can reach lengths of 4 centimeters. The apatite is often colorless or light yellow, and less commonly lilac or rose-red.

Left: Fluorapatite with orthoclase and chlorite, 8.1 cm tall
near Fiesch, Canton du Valais, Switzerland
John Lucking collection, Jeff Scovil photo

Upper right: Fluorapatite, 2.5 cm tall
Canton Uri, Switzerland
Luís Miguel Fernanadez Burillo specimen, Jeff Scovil photo

Lower right: Parallel intergrown apatite crystals, 2.1 cm tall
Collected in 1938
Val Casatscha, Switzerland
Marco Pellecchia collection, Roberto Appiani photo

A number of localities in the Saint Gotthard Massif are important producers of apatite specimens. Superb apatite crystals, as well as hematite, the adularia variety of orthoclase, and the pericline variety of albite, have been found in the **Gotthard Pass** (Gotthard Hospice) region. Crystals associated with rutilated quartz and hematite are abundant at Gotthard Pass. Choice crystals to 7 centimeters occurred in the gneiss of **La Fibbia** (Mt. Fibbia, Poncione della Fibbia) in association with muscovite, adularia, and pericline. An unusual find of apatite with phenakite occurred east of **Lago di Lucendro** (*lago* is the Italian term for "lake"). Tabular, enamel-like crystals, to 6 centimeters, some with titanite, came from the **Lago della Sella** region. (Yale University's Peabody Museum of Natural History houses several Sella apatites that were obtained from Kenngott in 1864.) The **Val Casatscha** region produced beautiful, clear, lilac-colored, prismatic crystals associated with quartz and pericline. Finally, excellent crystals with axinite, adularia, pericline, and chlorite are found in the granodiorites of **Val Cristallina**.

In the Aar Massif, outstanding lilac-colored crystals with quartz and adularia were found near the **Rhônegletscher** (*gletscher* is the German term for "glacier") and the **Grosstal** (Urserental). Fine complex crystals with titanite, adularia, quartz, and pericline came from the **Fieschergletscher** region. Very fine crystals with fluorite and adularia were found at **Schöllenen Gorge**. Superb tabular, rose-red crystals with brookite, albite, and quartz were found at **Reuss Gorge**, southwest of Amsteg (the western and eastern slopes are known as "Intschi Tobel" and "Rieder Tobel," respectively).

An historically significant locality in the Aar Massif is the upper **Val Giuv**, where apatite is associated with milarite (crystals to 3 centimeters), quartz, adularia, calcite, actinolite (variety byssolite), and chlorite. Val Giuv is the type locality for milarite.

KNAPPENWAND, AUSTRIA

The Knappenwand ("Miner's Wall") in the Untersulzbachtal (Hohe Tauern, Salzburg) is famous for producing some of the world's best epidote crystals. But it is also recognized as the source of the finest water-clear, beautifully formed apatite crystals ever found. These were determined to be fluorapatite by Adolphe Carnot in 1896. More recent analyses done by the Natural History Museum of Vienna (Seeman 1986) have shown the apatite to contain 1.7 to 2.6 weight percent fluorine and 0.07 weight percent chlorine.

Many of these crystals exhibit acicular actinolite ("byssolite" or "amianthus") inclusions. They are tabular in habit, predominantly showing the simple hexagonal prism and are terminated by the basal pinacoid, which is modified by very small pyramids up to the third order. The apatite crystals range in size from a few millimeters to 8 centimeters and occur either as individual crystals (very rarely as twins to 0.6 centimeters) or in intergrown parallel groups.

Especially striking specimens are produced when the limpid apatite crystals are associated with lustrous green epidote and brilliant white albite crystals. Due to the great value and high demand for these specimens, a number of matrix specimens have been faked. Either the apatite crystals have been glued onto epidote or have been inserted, along with the epidotes, into thick tufts of byssolite. It is wise to be mindful of the possibility that specimens may have been constructed, as a fair number do exist.

The Knappenwand deposit is hosted in epidote amphibolites ("prasinites") of the so-called Knappenwandmulde series in which many Alpine clefts occur. Those clefts that contain the wonderful epidote and apatite mineralization are usually oriented perpendicular to the epidote amphibolite beds. They contain the ubiquitous byssolite, which is one of the first cleft minerals to have formed, and hence it is found as inclusions in later-stage minerals and is especially noticeable in apatite.

In terms of history, copper mining activity in the Untersulzbachtal region has been recorded since the early 1500s. But the Knappenwand was not formally discovered until 1865. Alois Wurnitsch, an accomplished mountain guide and mineral collector, was the discoverer. After finding a number of single, loose epidote crystals to 7 centimeters and a cleft filled with clusters of epidote associated with apatite, calcite, and byssolite, Wurnitsch sold several specimens to his friend Andreas Bergmann, a part time mineral dealer in Innsbruck, Austria. The two men kept the locality a secret for several years. Bergmann sent a group of specimens to Professor Victor R. von Zepharovich, of the University of Prague, for scientific (mainly goniometric) analysis. Eventually, Zepharovich published the first description of Knappenwand mineral species in an 1869 paper entitled "New Mineral Finds in Salzburg."

Bergmann obtained a lease on the locality and, with his assistant Peter Trojer, began mining. Many fine specimens were found, and a selection of these were acquired by the Natural History Museum of Vienna. Aristides Brezina (1848–1909), curator of the Imperial Mineral Cabinet in Vienna, later joined Bergmann. Together they collected a huge number of specimens not only of epidote but also of apatite, titanite, calcite, and adularia. Brezina published a description of these species in a paper of 1871.

Above: Fluorapatite and actinolite (var. byssolite), 1.6 cm wide
Knappenwand, Untersulzbachtal (Hohe Tauren, Salzburg), Austria
John Rakovan collection, Jeff Scovil photo

Upper right: A view from inside a cleft at the Knappenwand
Hugo Ullhofen photo, circa 1930s, courtesy of Erwin Burgsteiner

Lower right: A circa 1930s photo showing Peter Trojer (standing) with Karl Wurnitsch (seated), the son of Alois Wurnitsch
Photo courtesy of Erwin Burgsteiner

Fluorapatite , 8.1 cm tall
Murchnerkar, Zillertal, Austria
Marty Zinn collection, Jeff Scovil photo

Bergmann became quite wealthy by selling Knappenwand specimens across Europe to collectors, dealers, universities, and museums alike. Knappenwand epidote and apatite became world famous and are justly considered jewels of the mineral kingdom.

Over the years, the Knappenwand locality has been intermittently worked for mineral specimens by a number of different lessors including the Natural History Museum of Vienna. The current leaseholder is the Zukunftkollegium Neukirchen.

ZILLERTAL, AUSTRIA

Zillertal is the largest valley that branches off the Inn Valley in the Austrian North Tyrol. Many fine apatite specimens have been found in the Alpine clefts scattered throughout the Zillertal. The clefts occur in gneisses, granites, and granodiorites. In his 1889 doctoral thesis John Leigh Hoskyns-Abrahall determined, among other things, that the apatite from the Zillertal is fluorapatite.

Some of the specific apatite localities in the Zillertal are **Floitental**, **Greiner**, **Rotkopf**, **Schwarzenstein**, and **Murchnerkar**. Crystals from these sites range in size from microscopic to 15 centimeters wide. They vary in habit from prismatic and columnar to the most frequently occurring thick tabular. The crystals generally exhibit white, glossy, and opaque to somewhat translucent faces. Common associations are chlorite, calcite, albite, and adularia.

Reinhard Brauns' (1912) magnum opus *The Mineral Kingdom* includes a chromolithograph of three Zillertal apatite specimens (see page 2, this volume). The first two crystals are from Schwarzenstein, and the third is from Floitental. A particularly fine suite of well documented Zillertal apatite specimens are preserved in the Yale University Peabody Museum of Natural History collection. These date back to 1872 and were obtained from various sources including L. Stadtmuller, Bohm & Wiedemann, Krantz, and the Austrian K.K. Mineralien Cabinet.

Apatite Valleys: Localities in Northern Pakistan

by Dudley Blauwet

For more than 35 years, the Islamic Republic of Pakistan has been a source of spectacular apatite specimens. While the species occurs in deposits in many areas of the country, the primary specimen-producing areas are in the Gilgit-Baltistān semi-autonomous region of northern Pakistan. The mountainous area that is home to these localities had been known as "Northern Areas" since 1970, when the Gilgit Agency and Baltistān district (including the states of Hunza and Nagar), were merged under that name. However in 2009, the administrative unit was granted limited self-rule and its name was changed to "Gilgit-Baltistān." The approximately 73,000 square kilometer (28,000 square mile) territory consists of seven districts: Astor, Diamīr, Ghanche, Ghizar, Gilgit, Skārdu, and Hunza-Nagar. The region borders the Pakistani provinces of Khyber Pakhtunkhwa and Azad Kashmir; the Wākhān region, Badakhshān, Afghanistan; the Indian state of Jammu and Kashmīr; and Xinjiang Uygur Zizhiqu in China. Gilgit-Baltistān is home to nearly one million people.

The most important apatite-producing areas in Gilgit-Baltistān are the world famous pegmatites of Chumar Bakhoor (Chomar Bakor) in the Hunza-Nagar District; the pegmatites near Shengus, Balachi, Saichis, Khargulook, and Sabsar, along the Indus River in the Skardu District; the Alpine cleft deposits around Alchori, Hashupi, and Chhurka in the Shigar Valley and Tormik (Tormiq) on the southwestern side of the Haramosh Range; and the pegmatites in the Braldu and Basho river valleys. The latter two series of localities, like those along the Indus, are in the Skardu District. As is common for apatite-bearing granitic pegmatites all of those discussed belong to the LCT family (see page 21).

CHUMAR BAKHOOR AND INNA NALA

In an extensive interview in fall 2012, Master Gohar Hayat, a school teacher who was one of the original discoverers of the **Chumar Bakhoor** deposit recalled that in the late 1970s a hunter from the village of Sumayar Nagar would regularly return to the village with fragments of aquamarine, but he kept their source a secret, claiming instead that he had found them near the 7,200-meter-high Diran Peak. Then in 1982 Hayat and an associate, Mohammed Shafi, ventured into the mountains where they encountered two miners who had small samples of aquamarine, but they too lied about the source. Hayat observed the shallow diggings in the distance, and he and Shafi climbed to the site. They found fragments of aquamarine and then returned to Sumayar to inform the village as to the whereabouts of the occurrence.

The next day, some 100 men climbed to the pegmatite, located at an elevation of over 4,600 meters on the western slope of Chumar Bakhoor Peak, and started to work. At first ropes were attached to the specimens, and horses pulled them out of the veins. But by the late 1980s the miners had started using explosives in a chaotic rush to collect specimens. Huge quantities of aquamarine and a lesser number of apatite and fluorite specimens were recovered although many were damaged through the extraction process.

The author and his then-business partner, Amir Ullah, first visited Sumayar in 1989. Isolated from the Karakoram Highway by the Hunza River, Nagar was also culturally insular. It seemed a separate country within Pakistan. The children were dirty, and most of the houses made simply of stacked stone. At that time, the descent from the Karakoram Highway down a dirt jeep trail to the rickety bridge over the Hunza River was dangerous. And the climb up the opposite side of the valley was even more so. Although the locals spoke Burushaski like their Hunzakut neighbors, they did not have the outward friendliness and did not offer the same customary cup of tea to guests.

In those early days, the men would cover the floors of the house with specimens to show them. The specimens mostly consisted of aquamarine, but there were some fluorite and apatite specimens as well. Most were large and quite damaged. Initially the locals were calling apatite "morganite," not realizing that they were a different species. Prices were commonly based on size, rather than aesthetics.

Countless specimens, along with some lives were lost at the mining sites in the early 1990s. By the time the author visited the locality on September 1, 1994, a miners' union had been formed. The union assigned groups of miners to specific sections of the deposit to avoid hazards such as rock falls from uncontrolled blasts that had been killing or maiming miners working in areas below the now-scheduled blasts. In a single year in the early 1990s, rock falls or snow avalanches had killed seven miners, including some working well into the winter season, when snow on the mountain was deep.

The union had also established a closing date of November 1st, but in many years mining ended instead in mid- to late October with the first heavy winter snow, and in 2011 the season closed several weeks into November as the snows came late that year. The union also established a system whereby a portion of the production was to go to non-mining families in the village (Fida Hussain personal communication 1994). That program transformed the villages far more than any aid group could have. Today, a second road has been built between Sumayar and Askurdas; every child has a freshly scrubbed face and a clean school uniform. Master Hayat said proudly that every child in the village, boys and girls, attends school, where they all attend rudimentary English classes along with their normal schooling. New houses constructed of concrete blocks have been built throughout the village. Specimen mining has eradicated poverty in the area.

Above: Fluorapatite with albite and muscovite, 3 cm tall
Chumar Bakhoor, Hunza-Nagar District, Gilgit-Baltistān, Pakistan
Steve Maslansky collection, Jeff Scovil photo

Facing page: The miners' camp at Chumar Bakhoor
Galibert Olivier photo, April 2010

In 1994, a rough miner's camp consisting of about a dozen large heavy canvas tents was in place on a knoll at about 4,630 meters (15,200 feet). The trail to the camp started at the Sumayar power station, which was at about 2,400 meters (8,000 feet). The lowest exposure of the main pegmatite was about 60 meters above and to the right of the camp. It drifted 10 or so degrees up and southeast, toward Diran Peak. At that point it had been worked for about 600 meters, and the tunnels were no more than 18 meters deep, because at that altitude and with the blasts sucking all of the oxygen from the unventilated tunnels, it was difficult to go much deeper. Many small, exploratory tunnels had been dug into the smaller pegmatites, which are for the most part above the main deposit.

At that time the author had a limited partnership with Sher Ali, a man from Hunza who was working at Chumar Bakhoor. The two stayed in one of the tents with seven or eight others one night in September 1994. The head of a freshly killed yak

Upper left: Fluorapatite on muscovite, 9.1 cm wide
Chumar Bakhoor, Hunza-Nagar, Gilgit-Baltistān, Pakistan
Stuart Wilensky specimen
Jeff Scovil photo

Lower left: Fluorapatite with muscovite, 3.5 cm tall
Chumar Bakhoor, Hunza-Nagar, Gilgit-Baltistān, Pakistan
Shirley Fiske collection
Jeff Scovil photo

Facing page, upper: Fluorapatite on muscovite, 8.8 cm tall
Inna Nala, Hunza-Nagar
Gilgit-Baltistān, Pakistan
Steve Smale collection
Jeff Scovil photo

Facing page, lower: Map of the Chumar Bakhoor and Inna Nala areas
William W. Besse cartographer

was dangling from a pole inside the tent. They ate qorba, a thick, whole wheat, unleavened bread which could be kept for weeks and then made soft enough to eat by dipping into hot tea. At that altitude, even the standard lentils are difficult to cook properly as the boiling point of water is well below that at sea level.

The sparse diet gave the conniving Sher Ali the idea that the miners might be willing to barter meat for mineral specimens, particularly once into the mining season when they had little money in their pockets. So Sher Ali rented a nearly new Mercedes bus, purchased a herd of about 110 goats, loaded them on the bus, and he and his son drove them (the goats) to Sumayar and then herded them to the miner's camp, where they grazed and were traded off for various specimens.

Sher Ali recalled that late in the summer of 1992, he was working at the deposit when a cavity had been opened that contained a wonderful specimen consisting of aquamarine along with two large apatite crystals, but the specimen was obliterated when the miners set a charge to extract it. Countless specimens fell to similar fates. The specimens that made it out intact were often those that were partly buried in blades of muscovite, as these pieces were often lightly attached to the granite and could be removed with hand tools. Some very large specimens were also removed intact and brought down to Sumayar, where by the mid-1990s, large saws were being used to safely reduce the matrix. Dealers in Peshawar further pared down many of these specimens before offering them to the western dealers who were frequenting Peshawar at that time.

The apatite commonly occurs on matrix of coarse muscovite, with blades of muscovite often piercing, burying, and /or obscuring the crystals, which are almost always tabular and can reach 5 centimeters in diameter. The very best crystals from the locality are rich pink to almost red-pink in color and are quite lustrous. The most desirable of these specimens are the rare and colorful combinations of apatite, aquamarine, and ideally pink or green fluorite all occurring on muscovite and /or albite. Occasionally apatite forms directly on an aquamarine crystal, but these are almost always smaller apatite crystals, rarely over 2.5 centimeters in diameter and more likely to be 1.5 centimeters or less.

The best single crystals of apatite from Chumar Bakhoor arguably came out in 1995 and 1996 although the author purchased an excellent miniature of apatite on muscovite in Sumayar in early November of 2001 (pictured on the cover of this volume). The top combination pieces are probably those that were recovered from the 1990s into the mid-2000s. More recent production is not of the same caliber as that from a decade ago. Since the mid-2000s, apatite production has decreased as the earlier worked portions of the veins are exhausted, and miners focus their efforts higher up on the western slope of Chumar Bakhoor. While production has decreased, prices have increased, and exorbitant rates are being asked for damaged, heavily cracked pieces, or excessively dirty specimens. In fact many specimens arriving from Chumar Bakhoor are coated and impregnated with rust, and safely removing the millimeters-thick layer of rust from apatite and porous muscovite without etching or damaging the luster on the apatite can be difficult.

The Chumar Bakhoor deposit is unusual in that, in addition to muscovite, twinned albite (variety pericline) crystals, much more typical of Alpine-type deposits than of pegmatites, commonly form the matrix for the later-stage deposition of beryl, apatite, and/or fluorite. Apatite crystals on twinned albite crystals to 2.5 centimeters, sometimes along with small rosettes of muscovite, have been found there. Microcline is uncommon at Chumar Bakhoor, and orthoclase (variety adularia) is even rarer, although crystals exceeding 7 centimeters in diameter have been reported.

Sumayar is the hub of the mineral market in the Hunza-Nagar district. In the late fall, owners of the shops and houses warm themselves in the sun while waiting for buyers to arrive. A few dealers also live down the road in Askur das Nagar (locally Askurdas). Three or four kilometers roughly southwest of Askurdas is the village of Phakker Nagar (locally Fikhar or Fiqhar), located on the south side of the **Inna Nala** (*nala* is the Urdu word for "valley"). Askurdas and Fikhar are connected by a jeep trail that traverses a steep barren slope.

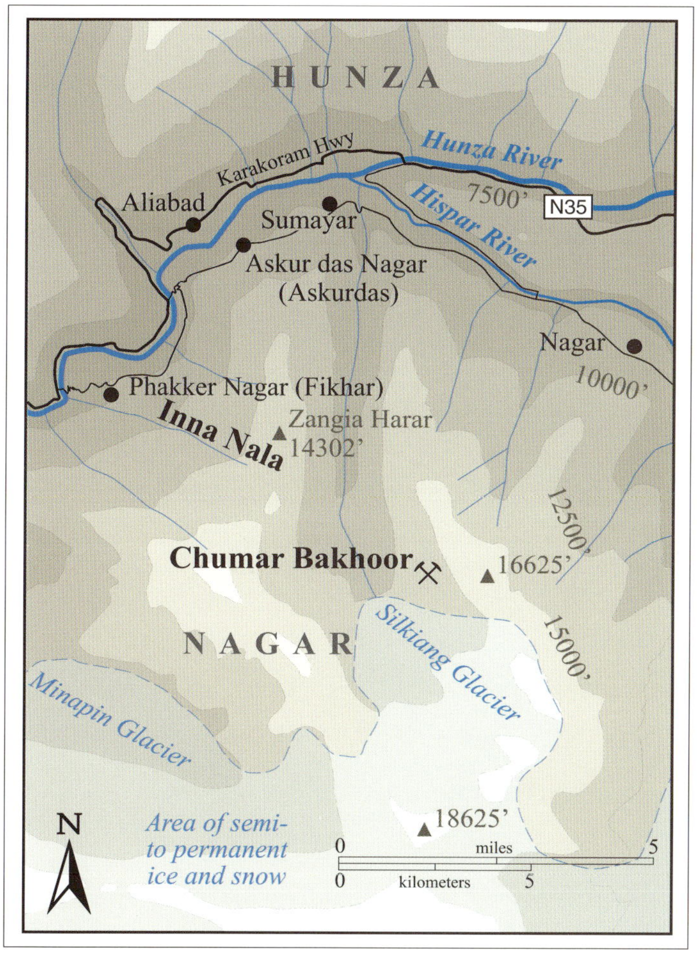

In the early 1990s a number of specimens of pink apatite were found in a small outcrop toward the top of the steep, narrow Inna Nala. Apatite crystals from this locality sport elongated prisms and are thus easily distinguishable from their invariably tabular counterparts from Chumar Bakhoor. The muscovite from the Inna Nala is also much smaller (usually 5 to 10 millimeters in diameter) and is densely compacted. Thus the apatite crystals are better exposed.

Inna Nala apatite is in general further distinguished by its color—more pink than red—and is commonly more translucent or transparent than crystals from Chumar Bakhoor. The small amount of aquamarine crystals that have been found at Inna Nala tend to be long, thin prisms, which are more transparent than those from Chumar. Finally, most of the Inna Nala specimens came out of the pockets without a coating of rust and are thus more lustrous.

One of the best specimens to come from Fikhar (pictured above) is 8.8 centimeters long and is composed of 6 gemmy, lustrous, pink apatite crystals, which are just over 2.5 centimeters long. They are scattered on a matrix of small muscovite crystals. The specimen was recovered in 1992.

Sadly, by about 1994, the Inna Nala locality had closed.

Apparently rocks from the mining operation were dropping on to the wheat fields below. The residents who owned the fields protested, and the mining operation ceased.

THE INDUS RIVER VALLEY

At 3,180 kilometers (1,980 miles) long, the Indus River is the longest and most important river in Pakistan. It originates in the Tibetan Plateau, runs the length of Pakistan, and ultimately moves roughly 207 cubic kilometers (50 cubic miles) of fresh water into the Arabian Sea each year. The Indus is vital to Pakistan's economy, and judging by the archeological finds along the river, some of which date back thousands of years, such has been the case for a long time. The sediments that the Indus has deposited into the Arabian Sea document the natural changes in the path and shape of the river over many millennia. Those changes, attributed both to nature and humans, continue to this day. The villages and side valleys along the Indus are natural gateways to the rich mineral deposits running through the daunting Karakoram Mountain Range through which the Indus cuts a path through Gilgit-Baltistān.

The Gilgit-Skardu Road parallels the Indus River, and the first notable occurrence along that road is an Alpine cleft deposit at the **Shuut Nala** in the Haramosh Range. The mine has seen limited production since the early 1980s, when it produced twinned epidote crystals. But in 2007 the author purchased a white, 2-centimeter-long, prismatic, gray-pink apatite crystal with oriented silky inclusions that produced a strong cat's-eye effect.

That anomalous find notwithstanding, the most significant

apatite deposits in the Indus Valley are pegmatites, particularly those concentrated around the village of **Shengus**. These pegmatites produce a variety of apatite colors, from pink to gray-green to colorless as well as a handful of blue crystals. There was even a limited find of small pastel, bicolor (pink-lavender to yellow-green), prismatic apatite crystals to 1.5 centimeters a few kilometers up the Indus at **Saichis** in the fall of 2007. Virtually all of the apatite has been found in pockets in pegmatites that also contain microcline or albite and occasionally groups of muscovite. Shengus area apatite is variously associated with elbaite, beryl (blue and colorless) garnet, topaz, schorl, and quartz.

A short distance up and on the opposite side of the Indus is the high, steep **Bulache** (Bulachi, Bolachi, Drot) Nala. The villages and the mines in this valley are located on some of the more shallow slopes high above the river. A notable apatite lot was recovered in this valley in the fall of 2010 from the **Shaukot Mine**. The best specimens consist of pink to pink-red, tabular apatite crystals to slightly more than 5 centimeters in diameter, most of which were found on microcline. Thick squat prismatic crystals to 3 centimeters were also found. Another discovery in the valley, this one in the mid-1990s, produced several aesthetic specimens consisting of multiple, tabular pink apatite crystals, to 2.5 centimeters on albite. The specimens range from 8 to 17 centimeters across.

In June 2011, the author purchased an excellent specimen, which originated in the area and consists of three dark green elbaite crystals with a number of 1- to 1.5-centimeter-wide, bright pink apatite crystals attached to them. And in the fall of the same year, the Shaukot Mine again yielded a few notable specimens of red-pink, prismatic, 2.5-centimeter-long apatite crystals on well-crystallized microcline. One of these specimens also consisted of several doubly-terminated aquamarine crystals along with three bright red-pink apatite prisms to 2.5 centimeters in length on a matrix of microcline.

Further upstream and on the same southwest side of the Indus, vast pegmatite belts streak the near vertical granite cliffs. The pegmatites at **Sabsar**, **Khargulook**, and **Saichis** are easily visible from the Gilgit-Skardu Road. On the Shengus side of the Indus, the deposits are located far above the road in various pegmatites stretching from above Shengus through Goochalay to Chhamachhu.

Combinations including beryl, topaz, tourmaline, and quartz and 1- to 1.5-centimeter-wide pink apatite crystals have been found at **Chhamachhu**, and in 2007 a colorless beryl crystal (variety goshenite) included with bright pink apatite crystals to 5 millimeters was found.

Nearby **Goochalay** is not generally known for apatite, but in 2007 the locality produced lustrous, brown euhedral examples of the species to 2 centimeters wide on well crystallized microcline as well as a significant group of lavender-colored apatite crystals on a roughly 22-centimeter-long quartz crystal. The latter specimen is remarkable in that the 0.5- to 1-centimeter-wide crystals form a long "snake," which stretches 10 centimeters along the quartz crystal.

This apatite-producing section of the Indus Valley is bookended 25 kilometers upstream from Shengus at **Stak Nala**, the famed tourmaline locality that has produced tens of thousands of bi-colored and multi-colored elbaite specimens. A handful of those multi-colored elbaite crystals have had tabular, pink apatite crystals to 2.5 centimeters attached to their terminations or to the beautiful albite matrix, which is characteristic of the locality. When present, those apatite crystals step the elbaite specimens up a notch.

Above: Fluorapatite, 10.8 cm tall
Tormik, Skārdu District, Gilgit-Baltistān, Pakistan
Neal Luppescu collection, Jeff Scovil photo

Facing page, left: Fluorapatite on topaz, 4 cm tall
Drot (near Bulache), Skārdu District, Gilgit-Baltistān, Pakistan
Irv Brown specimen, Jeff Scovil photo

Facing page, top right: Excavated pockets in a pegmatite in a cliff face above the Indus River
Alana McConnon photo, July 2010

Facing page, middle right: Jingle Truck on the Gilgit-Skārdu Road
Alana McConnon photo, July 2010

Facing page, bottom right: Mineral dealer Dave Bunk (left) crosses the Indus River via one of the many *gyari* aerial crossings

ALPINE-TYPE DEPOSITS

The Alpine cleft deposits in and around the Shigar Valley are also significant sources for Pakistani apatite. The first apatite-bearing Alpine-type deposit that bears mentioning—

the classic occurrence at **Tormik** (Tormiq)—is not in the Shigar Valley, but is situated in the Indus Valley above the village of Baghicha on the southwest side of the Haramosh Range, which separates the Indus from the Shigar and Basho valleys. Tormik was quite productive from the 1980s into the mid-1990s. The locality produced fine epidote, titanite, hematite, ilmenite, dolomite, siderite, hedenbergite, calcite, and gemmy etched quartz. And in the early 1990s, the first tabular apatite specimens were found.

A few specimens have been gemmy enough to be faceted into clean gemstones; some weighing several carats. Small biotite crystals are frequently observed on these crystals, the faces of which are quite dull, although the prism faces tend to be more lustrous than the terminal faces. The largest of the crystals are a bit wider than 5.5 centimeters and up to 2 centimeters thick. At times crystals exhibit an aggregate habit that is commonly found in Alpine cleft deposits. It consists of multiple tabular crystals that are closely intergrown in a parallel fashion but slightly offset from one another, creating a stair step-like topography. Many of the crystals were recovered loose, although some are attached to albite and/or are associated with other minerals such as calcite or actinolite (variety byssolite). Small, white tabular apatite crystals also occur at Tormik as a secondary species on the locality's mainstay epidote and titanite specimens.

Another variety of apatite produced by the locality exhibits a striking pearly luster. In fact, dozens of tabular, interconnected, pearly white, apatite crystals ranging in diameter from 5 millimeters to 2.5 centimeters form wonderful orientated rows, which can stretch across specimens exceeding 10 centimeters.

The locality has also produced a handful of unusual specimens over the past two decades. These consist of parallel series of yellow apatite crystals encased in blue-green serpentine "sheaths" that, when peeled away, may reveal six or more 1.5- to 2.5-centimeter-wide, rounded, tabular crystals with radial patterns emanating from the centers of their terminal faces. Their rounded outer surfaces and lozenge-like morphologies in their serpentine "wrappers" begged the author to dub these unique arrangements "Lifesaver" apatites. The yellow hue of much of the Tormik apatite is not stable: over time and especially with exposure to ultraviolet light, the color can fade to pale yellow, almost white.

Production at Tormik slowed by the mid- to late 1990s and was supplanted by that of the Alpine cleft deposits near **Alchori** (Alchuri) and **Hashupi** (Hashupa), both of which are

Fluorapatite from Nothern Pakistan's Alpine Clefts

Above: Fluorapatite with actinolite (variety byssolite), 3.2 cm tall
Alchori, Shigar Valley, Skārdu District, Gilgit-Baltistān
Dave Bunk Minerals specimen, Jeff Scovil photo

Right: Fluorapatite with clinozoisite, 5.8 cm tall
Alchori, Shigar Valley, Skārdu District, Gilgit-Baltistān
Mountain Minerals specimen, Jeff Scovil photo

Below: Fluorapatite with rutile, 4 cm wide
Alchori, Shigar Valley, Skārdu District, Gilgit-Baltistān
Francois Leitard specimen, Jeff Scovil photo

Facing page, left: A miner stands in front of an Alpine cleft at 3,000 meters on the west side of the valley north of Alchori; Dudley Blauwet photo

Facing page, upper right: Fluorapatite, 3.2 cm tall
Shigar Valley, Skārdu District, Gilgit-Baltistān,
Steve Smale collection, Jeff Scovil photo

Facing page, lower right: Fluorapatite on epidote, 2.4 cm wide
Shigar Valley, Skārdu District, Gilgit-Baltistān
Ryan Lay collection, Jeff Scovil photo

on the southwest side of the Masherbrum Range, which separates the Shigar and Braldu valleys. The villages of Alchori and Hashupi (Alchori is about 5 kilometers upriver from Hashupi) lie on the main road on the north side of the Shigar River and are at about 2,300 meters (7,600 feet) in elevation. Narrow valleys north of each of these villages pierce the Masherbrum Range. Miners have worked numerous Alpine cleft deposits exposed in these valleys. In the mid-2000s, the author visited one of the sites, which was located at 3,000 meters elevation on the west side of the valley near Alchori. In 2011 work had started at the Skinsar (the Balti term for "ibex") Mine. At an elevation of 4,500 meters, the mine is an 8-hour-long hike from Alchori.

While Alchori is more mineralogically varied and produces more and finer apatite specimens than does nearby Hashupi, numerous apatite specimens have emanated from both of these prolific localities. The very best apatite crystals from Alchori are transparent enough to be faceted into gemstones. Most are rather small (1 to 1.5 centimeters), although a handful are around 5 centimeters in diameter. Similar to those from Tormik, many of the larger apatite crystals exhibit stair step-like parallel intergrowths.

Alchori crystals tend to be pale pink to colorless, although yellow to light green, and a few pale blue crystals have been recovered. Their superb luster and association with other species, such as epidote, diopside, siderite, dolomite, clinozoisite, zoisite, orthoclase, hematite, and calcite, more than make up for whatever these crystals may lack in color. The locality has produced many miniature to small cabinet specimens consisting of 3 or 4 other mineral species alongside apatite. They are usually recovered on albite (sometimes

twinned), which is often covered with a thick layer of blue-green actinolite (variety byssolite) in the pockets. A 2004 pocket contained gemmy, yellow, tabular apatite crystals with bright red-bronze rutile needles on albite matrix. And in roughly the same period another unusual apatite occurrence was unearthed; this one consisted largely of single, thin, prismatic colorless crystals to 2 centimeters with white, hollow "tubes" down their centers—natural beads. A few of these hollow crystals were recovered on matrix, and a few of the "beads" are attached end-to-end.

Downriver, between Hashupi and the large village of Shigar, the village of **Chhurka** has also produced apatite of note. In the late 1990s, a small find of lustrous pink apatite crystals virtually identical to Alchori's best came from an Alpine-type cleft located above and northeast of the village at approximately 2,750 meters (9,000 feet). No further finds of apatite or any other mineral specimens were made from the locality until another lot of lustrous pink apatite crystals surfaced in 2004. Then both of these finds were trumped in the late fall of 2011 when a dealer bought two lots of pale pink apatite with quartz on matrix from miners who were working the site. The first lot contained one matrix specimen of a very lustrous nearly 3-centimeter-long, tabular, pale pink apatite crystal with quartz. The second lot, found a few days later, included several specimens of lustrous, 2.5- to 4-centimeter-wide, pale pink apatite with the stair step-like habit. The two largest specimens, among Pakistan's best Alpine cleft apatite, endured heart-breaking damage in the course of being mined.

Above: Fluorapatite with beryl (blue), 11.5 cm tall
near Baha, Braldu Valley, Skārdu District, Gilgit-Baltistān, Pakistan
Mountain Minerals specimen, Jeff Scovil photo

Facing page, left: Elbaite with fluorapatite and quartz, 7.6 cm tall
Chhaqpu, Braldu Valley, Skārdu District, Gilgit-Baltistān, Pakistan
Steve Maslansky collection, Jeff Scovil photo

Facing page, right: Fluorapatite with muscovite, quartz, and topaz, 7.3 cm tall
Nyet Bruk, Braldu Valley, Skārdu District, Gilgit-Baltistān, Pakistan
Marty Zinn collection, Jeff Scovil photo

THE BASHO AND BRALDU VALLEYS

In the early 1990s and sporadically throughout the next two decades, blocky, dark green to olive-green, prismatic apatite crystals were retrieved from Alpine clefts in the lower Basho (Basha) Valley, in the Haramosh Range near **Niaslo**. Although most are single crystals, some are remarkable in that they were found in association with semi-transparent, pale yellow-green orthoclase crystals. In the mid-1990s, this locality produced one of northern Pakistan's largest apatite crystals ever: a single, prismatic, dark olive-green crystal, 12.5 by 10 centimeters!

Like the Shigar and Indus, the Braldu River (gateway to the famous K2) provides an entry into otherwise inaccessible collecting areas high up in the Karakoram Mountains. The Braldu is one of the main tributaries of the Shigar, which in turn is a tributary of the Indus. The mined areas begin in the lower valley near **Baha** and **Teston** and extend upriver through Dassu, Nyet (Nit), Nyet Bruk, and finally to Apo Ali Gun,

Chhaqpu, and Ho Nala. The deposits in the Braldu Valley are not Alpine clefts but pegmatites. Braldu apatite is recovered from the prolific pegmatites, which produce much of the aquamarine for which Pakistan is famous.

Braldu apatite occurs in a variety of colors from smoky pink to yellow, green, purple, and rare blues and in both tabular and prismatic habits. It is often associated with more prominent crystals of aquamarine, schorl, or garnet. At **Teston**, in the lower Braldu Valley, a number of specimens consisting of overgrowths of red-brown garnet on long, thin prisms of pale pink apatite up to 4 centimeters long were found from the 1990s into the mid-2000s. A fist-sized specimen consisting of a dark blue aquamarine crystal with a purple-pink apatite crystal came out of the **Khandahar** pegmatite, northeast of the village of Baha, in the spring of 2011. Khandahar and the mines nearby have sporadically produced small pink to pale lavender apatite crystals for more than 15 years.

And while the occasional minor apatite specimen is known from a number of Braldu deposits, the pegmatites from **Nyet**, especially in the area of Shah Nassir Peak (a prominent dome that is nearly 5,000 meters high), have probably produced most of the apatite in the valley. A 2002 example from the locality consisted of a single tabular pink-gray apatite crystal, just over 5 centimeters across. And a nice discovery in 2011 was comprised of about 15 prismatic, moderately lustrous, medium pink crystals reaching lengths of 4 centimeters, some of which are associated with a small amount of muscovite. Unfortunately, much of the apatite produced here is dull, slightly etched, and has undesirable brown or gray tones.

Other notable Nyet finds include a bright yellow, prismatic crystal about 2 centimeters long on very fine albite that was collected in 2005; and some intense purple, tabular crystals, which are up to 1.5 centimeters that were recovered in 2007. Similar pink, purple, and lavender-colored crystals have been found on the other side of the valley at **Nyet Bruk**, particularly around the landmark Dapos "chimney." On both sides of the valley, apatite occurs in association with aquamarine, schorl, quartz, and very rarely with topaz.

Mining activity extends up the Braldu to Apo Ali Gun, which has over the years produced bright, tabular apatite in purple, muted pink-gray, and pale lavender that are sometimes associated with aquamarine and schorl. About 5 kilometers up and across the river is Ho Nala and the adjacent village of **Chhaqpu**, where similar pink-gray or lavender apatite crystals have been found. One of the more unusual discoveries at this locality was a specimen found in 2005 that consisted of a small, prismatic pink apatite and a prismatic pink beryl (variety morganite) along with several small, blue elbaite (variety indicolite) crystals, all on feldspar.

CONCLUSION

Pakistan's Gilgit-Baltistān region has produced thousands of specimens of apatite—some of the best in the world—over the past four decades. And there are undoubtedly still countless localities in the rugged mountains of the country waiting to be discovered.

Apatite from Nūristān, Afghanistan

by John Rakovan

Pakistan's Himalayan neighbor to the west is also home to numerous major pegmatite fields with hundreds of granitic pegmatites (Bariand and Poullen 1978; Bowersox and Chamberlin 1995). They are concentrated in the northeastern part of the country traditionally known as Nūristān, but which as of 2001 includes the provinces of Badakhshān, Kunar, Laghmān, Nangarhār, and Nūristān.

The geology of the area is complex with varied metamorphic and intrusive igneous rocks. The metamorphics are dominantly gneisses, mica schists, quartzites, and marbles, with igneous intrusions of granite, diorite and gabbros. The pegmatites are emplaced in all of the local lithologies—metamorphics, mafic, and felsic igneous suites—and are of variable size, ranging from veins of only 10 centimeters wide to bodies that are 40 meters in thickness. The pegmatites can be anywhere from a few meters to a few kilometers in length (Bariand and Poullen 1978). Alpine cleft type hydrothermal mineralization is also present in the area, and pockets containing adularia, chlorite, epidote, axinite, and apatite have been found.

Two pegmatite fields have been particularly significant for fluorapatite specimens: **Pēch Valley** (Darah-ye Pēch or Dara-i-Pech) pegmatite field, Chapah Darah District, Kunar Province, and the Nīlāw-Kōlam pegmatite field, Nūristān Province (formerly part of Laghmān Province); home to the famous **Mawī** pegmatite. Darah-ye Pēch is about 150 kilometers northeast of Kabul and about 40 kilometers from the border with Pakistan. The Mawī pegmatite is about 45 kilometers northwest of Pēch.

Both fields host LCT-type granitic pegmatites with

Upper right: Fluorapatite on microcline, 5.8 cm tall
Paprōk, Nūristān Province, Afghanistan
Marty Zinn collection, Jeff Scovil photo

Lower right: Fluorapatite on quartz, 9 cm tall
Nūristān Province, Afghanistan
The Collector's Edge, Inc. specimen, Jeff Scovil photo

Below: Mine adits along a cliff-side exposure of the Mawī pegmatite, Nīlāw-Kōlam pegmatite field, Afghanistan
Ikram Mohammad photo

Facing page, upper left: Fluorapatite, 2.5 cm wide
Khāsh district, Badakhshān Province, Afghanistan
Chris Stefano collection, Jeff Scovil photo

Facing page, upper right: Fluorapatite, 2 cm wide
Kunar Province, Afghanistan
Ryan Lay collection, Jeff Scovil photo

Facing page, lower: Fluorapatite, field of view ~9 cm tall
Mawī, Nīlāw-Kōlam pegmatite field, Nūristān Province
Arkenstone specimen, Joe Budd photograph.

mineralogies including spodumene, beryl, columbite-tantalite, quartz, albite, microcline, pollucite, muscovite, biotite, and fluorapatite (Orris and Bliss 2002). Deeply colored purple and blue fluorapatite crystals are much more common in the pegmatites of the Nūristān region than those in Pakistan. And other colors are found as well. Numerous specimens have reached the market over the years. One unusual and exceptional specimen is a 9-centimeter-long apatite "scepter" with a deep purple stem and light lavender cap that protrudes from a matrix of massive milky quartz (pictured right). The **Khāsh** district, Badakhshān Province, is also notable for producing royal blue clusters of fluorapatite.

The Nūristān region, particularly in what is Nūristān Province today, has seen tremendous conflict in the last several decades. The scene of some of the heaviest guerrilla fighting during the invasion and occupation of Afghanistan by Soviet forces from 1979–1989, it became a stronghold for Taliban rule. Consequently it has also seen some of the strongest conflict with American-led coalition troops in the last decade. This along with its inaccessibility and lack of infrastructure are among the reasons that mineral specimens from the region are far more scarcely seen on the international market than are those from Pakistan. Other impediments to mining and mineral export include the cost of explosives (US$1,200 for a box of dynamite at Paprōk in Nūristān), which are difficult to obtain (obviously anyone with explosives is viewed with suspicion), and the difficulties associated with moving anything from mine to market, as the numerous check points along the roads are often manned by corrupt officials who demand a portion of the shipment (Dudley Blauwet, personal communication).

China and Japan

by John Rakovan

Both China and Japan are home to several well-known apatite occurrences. Interestingly, however, the number of apatite-producing localities is small, and only one of the known occurrences are in granitic pegmatites. Given the size of China and the advances in resource development, future finds in both new and existing deposits are likely there. But new discoveries are less likely in Japan, and collectors need to rely on older material if they are to find specimens for their collections.

CHINA

Arguably the premier source for apatite specimens in China is **Mt. Xuebaoding** ("Snowy Treasure Peak"), Pingwu Xian (*xian* is the rough Chinese equivalent to a county), Mianyang Shi (*shi* is Chinese for "prefecture"), Sichuan (Ottens 2005; Liu 2006). Fluorapatite and a host of other minerals are found here in magma-derived hydrothermal veins, which are hosted by intercalating schists and marbles and in the intruding Pankou and Pukouling granites. Notable associated species include albite, beryl, cassiterite, fluorite, kësterite, muscovite, mushistonite, quartz, and scheelite, all of which are found with fluorapatite in beautiful specimens.

Fluorapatite crystals are typically equant to tabular and may exhibit numerous forms, but they are dominated by the first order prism and dipyramid and the basal pinacoid. Their color ranges from almost colorless to a rich raspberry pink, with most specimens exhibiting some intermediate shade of pink or violet-pink.

Southeastern China is exceptionally rich in hydrothermal polymetallic deposits, recently accounting for eighty five percent of the world's production of tungsten. Magmatic hydrothermal ferberite-hübnerite-sulfide-quartz veins and scheelite skarns provide the bulk of the reserves and production in the South China Tungsten Metallogenic Province. Although far less prolific, the many polymetallic deposits in this province are second to Mt. Xuebaoding in terms of production of fine apatite crystals. Located in the South China province, the important Nanling nonferrous metallogenic belt hosts the richest tungsten deposits in the world. This includes the well-known **Yaogangxian** deposit in Yizhang Xian, Chenzhou Shi, Hunan, the most important mine in the Nanling belt in terms of mineral specimen production (Ottens and Cook 2005; Ottens 2008; Ottens 2011). Fluorapatite crystals up to about 3 centimeters across are tabular to equant with a simple habit, usually exhibiting only the prism and pinacoid but sometimes the first order dipyramid. Crystals are colorless, white, gray to light green and are often in association with arsenopyrite, chalcopyrite, quartz, ferberite, and fluorite.

Similar deposits include the **Piaotang Mine**, Dayu Xian; the **Chongyi mining area**, Chongyi Xian, Ganzhou Shi; and

Above: View of 5,588 meter (18,333 foot) high Mt. Xuebaoding, "Snowy Treasure Peak," the highest mountain in China's Min Range and home to the Xuebaoding deposit

Facing page: Fluorapatite, 6.4 cm tall
Mt. Xuebaoding, Pingwu Xian, Mianyang Shi, Sichuan, China
Gail and Jim Spann collection, Jeff Scovil photo

Above: Fluorapatite and quartz, 6.6 cm tall
Yaogangxian Mine, Yizhang Xian, Chenzhou Shi, Hunan China
Steve Smale collection, Jeff Scovil photo

Below: Fluorapatite, 5 cm tall
Piaotang Mine, Dayu Xian, Jiangxi, China
Bert Ottens collection and photo

the **Xihuashan ore field**, Dayu Xian, all of which are in Jiangxi province. Apatite crystals from these deposits are generally similar to one another and those from Yaogangxian in habit, but they include a much wider variety of colors from colorless or white to cream, pink, green, gray-blue, and violet. Some particularly fine specimens in association with quartz and fluorite have been found in both Piaotang and Xihuashan, making both contenders for China's next most important apatite specimen locality. It is not uncommon for specimens from these occurrences to be mislabeled from "Yaogangxian," but the more colorful specimens are likely from Jiangxi.

Also in southeast China, about 320 kilometers northwest of Hong Kong, is the **Wutong Mine**, Liubao, Guangxi Zhuangu Zizhiqu (*Zizhiqu* is Chinese for "autonomous region"). A hydrothermal silver-zinc-lead deposit, the Wutong Mine has infrequently produced apatite specimens in association with quartz, pyrite, fluorite, and rhodochrosite. Crystals are typically less than a centimeter, equant to slightly elongate in morphology with a simple habit dominated by the prism and basal pinacoid. They are colorless, tan, golden, and violet, and color zoned tan-colorless, golden-violet.

Other notable Chinese occurrences of apatite include the topaz-elbaite bearing pegmatites in the **Gaoligong Shan** ("mountains"), Yunnan; the zinc-copper skarn of the **Lamo Mine**, Dachang tin field, Nandanzhan, Guangxi (Ottens 2008); and the iron-tin skarn of the **Huanggang Mine**, Kèshíkèténg Banner, Chifeng Shi, Nei Mongol ("Inner Mongolia") (Ottens 2012), but the number of specimens from these localities that have reached the world mineral market is inconsequential.

JAPAN

The most prolific and famous deposit for the production of apatite specimens in Japan is the **Ashio Mine**, Ashio-machi (*machi* is a suffix in Japanese which means "town"), Tochigi-ken (*ken* is Japanese for "prefecture"). Located about 110 kilometers northwest of Tokyo (Nambu 1970), the deposit was discovered in 1610 by a local resident, after which it was mined under the direct control of the Tokugawa shogunate (I.G.S.J. 1902). Until the Meiji period (which began in September 1868) this area was part of Shimotsuke province, and the old province designation may still be found on old specimen labels. Mining continued until 1973, and its high level of ore production, for most of its history, made Ashio one of Japan's most important copper deposits.

The mine also gained notoriety as the source of significant environmental damage: flooding, a result of upstream deforestation to fuel the mine, as well as the air and water pollution that the mine generated became flash points for nineteenth century Japanese environmentalists, who targeted the mine for reform and the area surrounding the mine for cleanup and restoration (Nimura 1998).

Subvolcanic hydrothermal mineralization occurs as polymetallic veins in the Ashio rhyolite. The deposit is mineralogically zoned and apatite is primarily found in the intermediate copper-zinc-arsenic zone along with quartz, chalcopyrite, and other ore and gangue minerals.

Fluorapatite crystals from Ashio are colorless to white to light brown, transparent to translucent. Many specimens exhibit zoned, pinkish orange and white-blue fluorescence under both short and long wave ultraviolet light. Crystal habit

Above: A modern view of a smoke stack at the Ashio copper mine, which was in production for more than 300 years and is today a museum
Ryoji Tanaka photo

Right: Fluorapatite, 9 cm tall
Ashio Mine, Ashio-machi, Tochigi-ken, Japan
Misao Sawada collection, Kiyoshi Kiikuni photo

Bottom: Fluorapatite, 6.4 cm wide
Ashio Mine, Ashio-machi, Tochigi-ken, Japan
Will Larson collection, Robert Weldon photo

Ashio Mine Ashio-machi Tochigi-ken Japan

燐灰石

varies from simple to complex; the basal pinacoid is dominant. The morphology is typically tabular, with crystals up to 3 centimeters across. More complex crystals exhibit small but numerous prism and dipyramidal forms (Wada 1904). Some pinacoid faces may show well developed hexagonal growth hillocks. Ashio apatite is crystallographically interesting because it exhibits lower than hexagonal symmetry and because numerous geometric domains with different optical orientations comprise the morphologic single crystals (Akizuki 1994). A fine apatite specimen from the Ashio Mine is pictured on Plate XXIb in Wada (1904). A second mine, **Kotaki**, exploits the same deposit as the Ashio Mine and has also produced apatite crystals. These are smaller, less than one centimeter, and are more equant than the typical Ashio crystals.

Chlorapatite and other compositions among the fluorine, hydroxyl, chlorine solid compositions are found at **Kurokura**, Yamakita-machi, Kanagawa-ken, in the southern part of the Kanto region. Crystals are almost always translucent to opaque white on their growth surfaces but transparent and colorless, straw-yellow or green-yellow in their interiors, a reflection of chemical zoning (Harada 1938). Their morphology is short to long prismatic, exhibiting first and second order prisms, basal pinacoid, and first and second order dipyramids. Although they have been found up to more than 4 centimeters in length, the majority of crystals are 1 centimeter or less. While most crystals are reported as "chlorapatite," the specific solution

Far Left: Fluorapatite, 4.7 cm tall
Kurokura, Yamakita-machi
Kanagawa-ken, Japan
Misao Sawada collection
Kiyoshi Kiikuni photo

Above: Fluorapatite, 3 cm wide
Kurokura, Yamakita-machi
Kanagawa-ken, Japan
Hiroyuki Imai collection and photo

Immediate left: Fluorapatite, kaolinite, and pyrite, 1.8 cm wide
Fubasami Mine, Imaichi-shi
Tochigi-ken, Japan
John Rakovan collection and photo

cannot be determined visually, and analyses of at least one reported "chlorapatite" crystal from Kurokura has proven to be fluorapatite (author's analysis). Kurokura apatite is found in association with quartz, chlorite, stilbite, and natrolite.

Euhedral crystals of apatite and pyrite embedded in kaolinite are found at the **Fubasami Mine**, Imaichi-shi (unlike Chinese, *shi* is used for "city" in Japanese), Tochigi-ken. Imaichi-shi is about 30 kilometers northeast of Ashio-machi. On March 20, 2006, Imaichi and Ashio were incorporated into the city of Nikkō, thus putting two of Japan's premier apatite localities (of very different geologic origin) in the same municipality. The deposit has been worked for more than sixty years and still produces apatite crystals today. Apatite from the locality is transparent, green-yellow in color, and often shows faint color zoning perpendicular to the *c*-axis. The crystals are simple tabular with the basal pinacoid dominant. Prism faces are highly lustrous and may show subtle but well developed striations and growth hillocks. Pinacoid faces are often pitted from the presence of a preexisting mineral. One large crystal has been observed that exhibits a thin central zone. It fluoresces orange-pink in its center and blue on its two exterior surfaces in long wave ultraviolet light.

Chlorapatite and other apatite supergroup minerals including hydroxylellestadite are found in a skarn deposit of the **Chichibu** Mining District, Chichibu-shi, Saitama-ken, just south of Tochigi-ken. Associated minerals include magnetite and pyrrhotite as well as andradite and diopside. Apatite forms short prismatic crystals with prism, pinacoid, and multiple dipyramidal forms exhibited. They are white to pale yellow and transparent to translucent. Crystals attaining 3 centimeters in length have been found, but they are usually less than 1 centimeter. Like samples from Kurokura, the chlorine content is variable and cannot be estimated visually, thus identification as "chorapatite" is suspect without chemical analysis.

One last noteworthy locality, although there was only one significant find of apatite recorded from there, is on Mukai-yama (*yama* means "mountain"), **Otome** Mining District, Yamanashi-shi, Yamanashi-ken (formerly Kai or Koshu province). What is referred to as the "Otome Mine" is actually a district comprised of several different mines that exploit hydrothermal quartz veins in the Shosenkyo Granite. Some of the other mines in the district are in fact in the neighboring city of Kofu. The Otome is famous for large Japan-law twinned quartz crystals; it was the original source of Japan's most notable specimens. It is also known for "reinite" (ferberite pseudomorphs after scheelite). In the 1880s a large pocket containing transparent yellow apatite was found. Wada (1904) reported crystals up to 9 centimeters in diameter, he also noted that chemical analyses made by H. Yoshida showed the specimen to be fluorapatite with minor chlorine.

ACKNOWLEDGEMENTS: *The author would like to thank Bert Ottens, Alfredo Petrov, and Safaa Yu for reviewing this manuscript and for the many discussions of China and Japan.*

Apatite Localities of Russia And Other Former Soviet Republics

by Dmitriy Belakovskiy, *Fersman Mineralogical Museum, Moscow*

The greatest concentration of apatite in Russia, and the world, is on the Kola Peninsula. Most of the Kola apatite localities are related to alkaline and ultramafic massifs including Khibiny, Lovozero, and Kovdor. Khibiny is one of the world's primary producers of apatite—the source for superphosphate fertilizers—although Kovdor is also a significant producer.

The story of the discovery of the Kola Peninsula apatite deposits is an interesting one. In 1921, Alexander E. Fersman, director of the Mineralogical Museum at the Russian Academy of Science (now the Fersman Mineralogical Museum), organized an expedition to **Khibiny**. The museum expedition was supported by the Commission on Investigations of Natural Productive Resources of Russia. This commission, a brainchild of Vladimir I. Vernadskiy, was strongly encouraged by the new Communist administration, which had an interest in surveying the country's resources, as little data on the mineralogy and geology of the Kola Peninsula had been collected by that time.

While investigating the mineralogy of Khibiny was one of Fersman's goals, he also planned to study the alkaline pegmatites as a potential new gemstone source. Practically no gem materials were found, although the potential for mineralogical discoveries was so great that Fersman dubbed the place "mineral heaven." Among the expedition's discoveries were outcrops of unusual apatite rocks at Rasvumchorr Mountain that were found by museum employee, mineralogist, and former Colonel in the Tsar's Army, Alexander N. Labuntsov. Labuntsov accurately estimated the economic importance of his find; still, a number of prominent geologists were skeptical. The most apatite-rich deposits known at the time were related to sedimentary formations—an economically viable apatite deposit in nepheline syenite was unusual and thought unlikely.

Upper: View of the Northern Lights over the Khibiny Massif Aleksander Chernucho photo, November 2008
Lower: A.E. Fersman and his team in the Khibiny Mountains, 1921

Above: Fluorapatite, crystals to 1 cm with acicular aegirine and red kentbrooksite on white orthoclase
Hackman Valley, Khibiny Massif, Russia
Private collection
Boris Kantor photo

Right: Apatite-nepheline rocks at the Kirovskiy Mine, Khibiny
Kola Peninsula, Russia
Viktor Levitskiy photo, July 2012

A year later, Labuntsov was back at Khibiny. He rappelled down a vertical cliff of apatite-nepheline rocks and, marking his climbing ropes, measured the thickness of the apatite bodies. Those measurements, along with chemical data on the phosphorus concentration of the apatite-nepheline bodies, proved that this was not just a big apatite deposit but the world's largest! Additional expeditions were quickly organized, and commercial mining for apatite commenced in 1929 at Yukspor Mountain, near the newly established mining town later named Kirovsk. The city of Apatity ("Apatites") was established in 1935 to support the mines. At present, after more than 80 years of mining, the deposit is still being worked via a number of huge open pits and underground mines. It remains the world's largest source of apatite.

The Khibiny massif is north of the Arctic Circle. It is picturesque and is a pleasant place to visit in the summertime, when it is sunny and the mosquitoes take a rare holiday. It is not at all pleasant, however, during the long, windy, snowy polar winters.

Geologically speaking, Khibiny is the world's largest circular alkaline massif. It is some 20 kilometers in diameter and consists of concentric zones of nepheline syenites. The series of iyolite-urites is emplaced in syenites in the middle part of the massif. The apatite-nepheline rocks are related to and alternate with iyolite-urites (photo above). The apatite forms spotty textures of sugary, light green to black aggregates. The black crystals derive their color from inclusions of organic matter. In these bodies, apatite never forms good crystals, but the ores can be attractive when they also show spots of black aegirine, brown titanite, red eudialyte, and dark green nepheline. The titanite content is often significant enough for it to be considered apatite-titanite-nepheline ore. Nepheline, titanite, and eudialyte in these rocks are always euhedral. The eudialyte forms nice crystals up to 5 centimeters. In terms of color, the best eudialyte occurs in apatite ores.

In addition to being a source of phosphorus, apatite is also a source of strontium, rare earth elements (REE), and fluorine, all of which are sporadically extracted from the ore. These apatite-nepheline rocks are also potential sources of aluminum (nepheline), titanium (titanite), and vanadium (aegirine); in fact the original plan called for 12 different elements to be extracted, and huge piles of ore are waiting to be processed.

Apatite is a common mineral throughout the Khibiny, but large, collectible crystals are rare. They are generally found in pegmatites and usually formed during the latter stages of mineralization. The best examples were recovered from cavities in a thin, feldspar-bearing pegmatite vein that cuts mosandrite-containing rocks. The vein is exposed in an underground mine (worked in the 1930s for thorium and rare earths) at Yukspor Mountain in the **Hackman Valley**. Transparent, light green, short prismatic crystals up to 2 centimeters with pinacoid terminations are found in association with astrophyllite on pale feldspar matrix. Most of the apatite crystals are included with acicular aegirine crystals. They are further associated with white natrolite and are dotted with small, red crystals of eudialyte-group minerals (photo above).

Chemically, the apatite in this vein is more enriched with strontium and rare earth elements than is the apatite from the ores and is locally referred to as "saamite" (for the indigenous Saami people who live in the area). This small vein produced a limited number of specimens.

Similar apatite-bearing rocks are found in the **Lovozero** circular alkaline massif. Lovozero is smaller than Khibiny but is still one of the largest massifs of its type. Serious geological investigations began in 1923, also by the museum expedition. The massif later became known for super-large niobium-tantalum orebodies related to enormous loparite-(Ce) concentrations. Apatite does not form large orebodies at Lovozero, but it is still a common mineral in syenites and pegmatites. Significant numbers of the same saamite variety of apatite, including crystals up to 20 by 8 centimeters, have been recovered from the natrolite core of **Pegmatite #47**, which is situated on the slope of Lepkhe-Nelm Mountain facing Lake Seidozero. Lepkhe-Nelm Mountain crystals are typically not as transparent as those from the Hackman Valley. They are bright yellow-green and are dominated by the prism

Right: Fluorapatite
17 cm wide
Panfilova varaka
Karelia Republic, Russia
Fersman Museum collection
(cat. no. 27982),
Natalia Pekova photo

Below: Fluorapatite (variety francolite), 8 cm
Kovdor, Kola Peninsula, Russia
Mikhail Bogomolov collection and photo

{100} and pinacoid {001} forms, which are on rare occasions modified by the dipyramid {101}. They also form radial aggregates up to 10 centimeters in diameter as well as some hoppered crystals. Saamite from Lepkhe-Nelm is enriched with up to 12 weight percent rare earth element oxides and up to 11 weight percent strontium oxide. These levels could be the result of micro-inclusions of another apatite-group mineral—belovite-(Ce) (Pekov 2001). Bright, yellow to green, prismatic crystals of belovite-(Ce) up to 5 by 1.5 centimeters are the best of the species found in the Lovozero Massif. They occur in an ussingite body at Malyi Punkaruaiv Mountain.

Blue to deep blue, pink or purple manganese-bearing apatite crystals up to 0.5 centimeters in natrolite are also known from Lovozero. These are found in pegmatites on **Lepkhe-Nelm**, **Kuivchorr**, and **Flora** mountains. In both Khibiny and Lovozero, carbonate-bearing fluorapatite (varieties francolite or staffelite) in different colors are widely distributed in weathered zones along the faults.

Fluorapatite is a secondary ore mineral at the **Kovdor** ultramafic carbonatite massif, which was discovered in the Kovdor River Basin on the southwestern portion of the Kola Peninsula in 1933. The deposits are and were mined for iron, phlogopite, and zirconium (baddeleyite). Fluorapatite is an accessory mineral in the ultramafic and alkaline rocks in the massif and commonly forms sugary, greenish aggregates with magnetite, phlogopite, and calcite (photo left). It sometimes forms parallel or radial crystal groups up to 1 meter in diameter. Long, prismatic crystals up to 15 centimeters occur in diopside. Probably the most attractive apatite specimens from Kovdor are of the "francolite" variety, which commonly forms as white, red-brown to orange crusts but sometimes as delicate greenish or turquoise-blue botryoidal aggregates in weathered zones along the faults. The blues and greens are caused by micro-inclusions of brochantite or malachite.

Apatite crystals are also found in the granitic pegmatites in the Karelia Republic, south of the Kola Peninsula. This area was long worked for muscovite crystals for use in windows. In the **Chupa** area, where pegmatites were mined for mica until only recently, apatite occurs as green-blue or blue (due to trace amounts of manganese) crystals up to 20 centimeters and in association with feldspar (photo above). The crystals are euhedral but are not transparent and are usually fractured.

Apatite is common in the form of phosphorite, carbon-bearing fluorapatite, throughout the rest of the European side of Russia and in the former Soviet states. Phosphorite forms concretions in sedimentary rocks. Economic deposits of phosphorite are common in the central regions of Russia. Many (the **Egorievskoe** deposit near Moscow, for example) contain great quantities of fossils (mostly ammonites), which are replaced or overgrown with apatite. The mineral is also

Above: *The Battle of Poltava*
Image courtesy of the Poltava Battle History Museum
Poltava, Ukraine

Left: Carbonate-bearing fluorapatite (phosphorite)
9 cm wide
Zhukovka, Orlovskaya Oblast', Russia
Fersman Museum collection (cat. no. 42939)
Natalia Pekova photo

Right: Fluorapatite, 4.6 cm wide
Dodo, sub-Polar Urals, Russia
Brian Kosnar specimen, Jeff Scovil photo

Facing page, upper: Fluorapatite with clinochlore, 5 cm wide
Ahmatovskaya Kop', near Zlatoust, Southern Urals, Russia
Fersman Museum collection (cat. no. 32864)
Natalia Pekova photo

Facing page, lower: Fluorapatite, 3 cm long
Snezhinsk, Chelyabinsk area, Southern Urals, Russia
Oleg Lopatkin specimen and photo

commonly found in irregular-shaped or spherical nodules.

One of the more interesting phosphorite occurrences is located near the town of **Kam'yanets'-Podil's'kiy** (Khmel'nyts'kaya Oblast'), Ukraine. At this locality, highly spherical phosphate nodules occur in clay layers. They have knobby surfaces and reach up to 30 centimeters in diameter, although they only average about 9 centimeters (inset photo above). Broken nodules show radial structures and quite often contain galena and sometimes sphalerite in their centers. In the places where the concretion layers have been reworked by fluvial processes, the nodules are naturally tumbled and develop a smooth, polished surface.

Once upon a time these concretions, which are found in a number of places in the region, were actually used as cannon balls! In a 1709 battle between Russia and Sweden that took place near the city of Poltava, Tsar Peter the Great ordered his troops to collect phosphorite nodules, and the Russians bombarded the Swedes with them. Did the concretion munitions clinch a victory for Russia? There is no indication in the written history, but the apatite-armed Russians did win the battle, which was an important to Russia's efforts to expand to the west and north.

Urals

There is a great diversity of apatite in the Ural Mountains. Beginning in the north, colorless (sometimes with a hint of green or blue), transparent, tabular crystals of apatite up to 15 centimeters occur in Alpine clefts in chlorite schists in the **Dodo** and **Puiva** deposits, about 90 kilometers from the village of Saranpaul', Khanty-Mansiyskiy Autonomous Okrug near Neroika Mountain in the sub-Polar Urals. These deposits were discovered in the 1930s by government geologists who were prospecting for piezo-optical quartz. The fluorapatite crystals have lustrous pinacoid faces, which show stepped growth patterns and hillocks. They are associated with clear quartz and are commonly sprinkled with small intergrowths of chlorite "leaves" (photo above). Other associated minerals include titanite and hematite. The specimens, mineral associations, and geologic characteristics are not unlike those seen in the Alpine clefts in Switzerland. Apatite crystals of similar origin have also been found at the **Rechnoe** quartz deposit in the Yuzhnyy ("Southern") Urals.

In the **Katsna Yama** pit, flat colorless apatite crystals up to one centimeter occur in association with dark-green, well-formed epidote (variety pushkinite) crystals. This pit is located near the city of Pervoural'sk in the Sredniy ("Middle") Urals. The **Izumrudnye kopi** (Emerald mines) about 100 kilometers east of Yekaterinburg (Sverdlovsk in Soviet times), is an important source for apatite specimens. Discovered in 1869, this series of deposits is famous for green beryl ("emerald"), chrysoberyl ("alexandrite"), and phenakite, but the apatite here is definitely interesting. It forms both short (up to 8-by-3 centimeters) and long (up to 5-by-.15 centimeters), prismatic crystals, which occur mostly in phlogopite-rich rocks that also bear emerald and alexandrite. True to its name, the apatite from here easily deceives collectors. Some green-toned varieties can be confused with pale-colored beryl while whitish or pitch-colored crystals could be wrongly taken as phenakite. Apatite from the Emerald Mines area is rarely colorless. The geology of the deposit is complex and the origin of the mineral associations is still a source of debate. A.E. Fersman described the occurence as "desilicified pegmatites."

Hydroxylapatite specimens are known from the **Shabry** metasomatic talc deposit near Yekaterinburg. It forms green-yellow crystals up to 2 centimeters in "noble" light green talc. The paragenesis of the deposit is similar to that of Holly Springs in the United States (see page 7).

Many fluorapatite occurrences in the Middle and Southern Urals are in skarn-type iron deposits. Examples include the **Lebyazhinskiy Mine** near the city of Nizhniy Tagil and **Gora Vysokaya** in the Middle Urals; **Kuybas** in the Chelyabinskaya Oblast', Southern Urals; and many others. The same type of deposit with the same type of apatite specimens extends south into Kazakhstan (the **Sarbayskoe** deposit, for example).

Apatite crystals associated with magnetite are sometimes large (exceeding 12 centimeters) but are rarely transparent. They are green-yellow, gray, brown, or appear red as a result of hematite inclusions. Combinations of a number of dipyramids terminate the crystals. A little chlorine content is typical for apatite from these deposits.

Apatite is also concentrated in titanomagnetite deposits related to magmatic gabbroids or syenite intrusions, such as **Kusinskoe** near Zlatoust, Southern Urals, or neighboring **Kacharskoe** in the Kustanay area, Qostanay Oblysy, Kazakhstan. Skarn-type processes are widespread in these deposits; thus, some of the apatite is metasomatic in origin. These processes probably explain the occurrence of chlorapatite at Kusinskoe.

Different deposits or occurrences, all considered skarns, are found north and west of the city of Zlatoust, Southern Urals (Chelyabinskaya Oblast'). Among these, the **Nazyamskie Mountains**, **Shishimskie Mountains**, **Akhmatovskaya Kop'** ("pit"), **Eremeevskaya Kop'**, and **Shishimskaya Kop'** have been known for a long time and were mined sporadically for magnetite ore. Akhmatovskaya Kop' is famous for classic specimens of andradite and clinochlore, but it has also produced nice apatite crystals found in association with barrel-like clinochlore crystals (photo upper right).

Another interesting occurrence in the Southern Urals was discovered relatively recently—toward the end of the 1990s—near the town of **Snezhinsk**, Chelyabinskaya Oblast'. Access to the area is tightly restricted, and as a result, little geologic data on the locality is available. Local collectors report a mica body consisting mostly of phlogopite that outcrops among serpentinites. The 2-meter thick mica body is exposed for a length of about 20 meters. The apatite is concentrated in nests near the contacts between the mica body and the serpentinites. Dark-green, well-shaped, and often slightly flattened apatite crystals (inset photo) to 15 centimeters occur in the phlogopite aggregate. Because of the shapes and colors of these crystals, they are easily confused with diopside. The apatite is often gemmy, and cabochons made from this material are an attractive yellow-green. Noticeable quantities of albite have also been found in these phlogopite aggregates.

Siberia

One of the world's most famous apatite localities is near the town of **Slyudyanka** at the southwestern termination of Lake Baikal (Irkutskaya Oblast'). Geological and mineralogical study of the area began in the final decade of the eighteenth century, if not earlier. Large, perfect crystals of apatite, phlogopite, diopside, scapolite, barian orthoclase, spinel, pargasite and many other minerals were found and described by Nikolay I. Koksharov in *Materialen zur Mineralogie Russlands* (1886).

The geology of the area is complex and diverse. Archaean granitoids intruded into highly metamorphosed sediments, including calcite and calcite-dolomite marbles, diopside-forsterite-phlogopite-calcite-dolomite calciphyres, gneisses and schists, and apatite-quartz-diopside rocks (thought to be metamorphosed phosphorites). Metasomatic processes formed the skarn-type phlogopite deposits, which were mined commercially from 1915 to about 1960.

Left: Fluorapatite, 7 cm tall
Akzhailyau, East Kazakhstan
KARP specimen, Jeff Scovil photo

Center left and middle: Pleochroic fluorapatite photographed with a polarizing filter with the polarization direction vertical and parallel to the *c*-axis (**left**) and horizontal, parallel to the *a*-axis (**middle**).
1.6 cm tall crystal (private collection) and 2.24-carat faceted stone (GIA collection cat. no. 13865)
Slyudyanka, Irkutskaya Oblast', Russia
Mark Mauthner photo.

Center right: Fluorapatite with hematite inclusions (brown), 3 cm tall
Seligdar Aldan, Oblast', Russia
Mikhail Bogomolov collection and photo

Facing page, upper left:
Fluorapatite in calcite, 4.6 cm tall
Slyudyanka, Irkutskaya Oblast', Russia
Marty Zinn collection, Jeff Scovil photo

Facing page, right: Fluorapatite and quartz
6.9 cm tall
Dashkezan, Caucasus Range, Azerbaijan
Collector's Edge specimen, Jeff Scovil photo

Facing page, lower left: Fluorapatite and calcite underground in Slyudyanka
Irkutskaya Oblast', Russia
Oleg Gritsenko photo

Giant, perfect crystals of phlogopite, scapolite (meionite) crystals up to 70 centimeters, diopside, barium-orthoclase, and pargasite are all found here, but the pearl of this deposit is apatite. It usually occurs in nests or veins of cream to yellow-orange calcite in diopside metasomatites. Euhedral, prismatic, doubly-terminated crystals can reach lengths of up to 1 meter, and 10- to 20-centimeter-long crystals are common. In most cases, the edges and faces of the crystals look as if they have been fused. Their color is usually blue or green-blue; they are rarely purple or have purple zones. Many crystals are opaque and included with calcite, but they are just as frequently gemmy and suitable for faceting. Some exhibit a cat's eye effect. Apatite from Slyudyanka is strongly dichroic, changing from pale to dark blue in polarized light (photo sequence above). It fluoresces under ultraviolet light. Small crystals of a deeper blue apatite are found in association with lazurite at the nearby Malaya Bystraya River.

Apatite crystals as large as 25 centimeters are also found in **Emel'dzhak**, **Snezhnoe**, and other phlogopite deposits around the city of Aldan in southern Yakutia Oblast'. The geology and mineral assemblages in these deposits are similar to those near Slyudyanka as well as in the Central Metasedimentary Belt, which cuts across Québec and Ontario, Canada (see page 38). There is, however, no blue apatite among the Aldan finds; crystals are dark green, like diopside, some have brown sections (photo above, right). Completely brown crystals (a result of hematite inclusions) at least as large as 30 centimeters are known from the **Seligdar** deposit, 30 kilometers south of Aldan.

The origin of the Seligdar deposit is similar to others around Aldan, but the mineral paragenesis is a source of debate.

Decent specimens of dark blue manganese-rich apatite frozen in albite matrix are known from a number of pegmatites in the **East Sayan Mountains**, west of Lake Baikal. Finally, vast apatite resources were discovered in **Maimecha-Kotuy** province about 600 kilometers east of Noril'sk and south of the Taymyr Peninsula in the 1940s. More than 15 ultramafic-carbonatite massifs contain orebodies not unlike those at Kovdor and Khibiny. These apatite orebodies are estimated to be comparable to those at Khibiny, but their location is well above the Arctic Circle, where there is simply no infrastructure; thus, exploration is problematic. No specimens of note have been recovered, but that is not to say that they will not be.

OTHER FORMER SOVIET UNION REPUBLICS

The **Dashkesan** (Daşkəsən) iron deposit is located 35 kilometers southwest of the city of Gyandzha (Kirovabad in Soviet times), in the small Caucasus Range in Azerbaijan. Dashkesan is a classic skarn deposit. The iron ores were known and have been mined by local artisans since ancient times. In the nineteenth century, the cobalt ores (cobaltite) were exploited as were the nearby alunite deposits. Commercial mining for iron started in 1954 and continues today. The Dashkesan deposit has produced excellent specimens of drusy apatite. Prismatic crystals up to 8 centimeters long are gray-green and have internal zones, which are usually a bit darker green. Pinacoid terminations are always modified by one or more dipyramids.

Single crystals or groups are found in druses with modified dodecahedral crystals of magnetite as well as with epidote, quartz, calcite, chalcopyrite, andradite, and pyrite. The mineralization is in vugs in hard andradite-epidote skarn rocks. As with apatite from other iron skarn deposits, Dashkesan apatite typically contains some amount of chlorine but never enough to make it chlorapatite.

In the Republic of Georgia, flattened apatite crystals up to 2 centimeters were found in quartz-calcite-dolomite veins in **Madne-tskali Ravine**, Verhnyay Racha.

Seal bones overgrown with apatite (alteration pseudomorphs?) have been found in miocenic sedimentary iron ore deposits (famous for an excellent vivianite specimens) near the city of **Kerch** on the Crimean Peninsula, Ukraine (photo back cover).

Right: View of Akzhailyau, East Kazakhstan, Kazakhstan
Viktor Ponomarenko photo, July 2006

Facing page: Fluorapatite, fluorite, and bertrandite, 12 cm long
Akchatau, Qaraghandy Oblysy, Kazakhstan
Marty Zinn collection
Jeff Scovil photo

Kazakhstan is home to a number of interesting apatite localities (in addition to those mentioned above). Most are related to granitic pegmatites. An occurrence in the **Akzhailyau Mountains**, Semipalatinsk (Semey) region, East Kazakhstan, is one of the most amazing. Quartz-feldspar pegmatites containing tourmaline, apatite, and fluorite (photo facing page) were found here in 1946, when geologists were mapping the area.

Hundreds of pegmatites are related to different phases of the Permian Akzhailyau granitic pluton. Pegmatites are located in both the granites and in the volcanic-sedimentary Paleozoic rocks that the granites intruded. The pegmatites reach up to 50 meters in length and up to 10 meters in thickness.

Pockets up to 100 cubic meters have been found. Large, zoned, smoky quartz crystals are common in those pockets as are euhedral crystals of other pegmatite minerals including microcline, albite, orthoclase (variety adularia), a range of colored quartz varieties, muscovite, biotite, allanite-(Y), monazite-(Ce), zircon, schorl, epidote, fluorite, and apatite. The latter of these forms prismatic crystals up to 6 centimeters with pinacoid terminations modified by dipyramids.

Akzhailyau apatite is remarkable in that it exhibits a strong alexandrite effect: crystals change from yellow-green in daylight to pink in incandescent tungsten light. This phenomenon is most likely caused by the presence of rare earth elements (total REE content 1.71 weight percent). Many crystals are transparent and are suitable for faceting; indeed, faceted stones weighing as much as 15 carats have been cut from Akzhailyau rough. The pegmatite that produced most of the specimens is mined out. In 2006 the Republic of Kazakhstan set the massif aside as a nature reserve.

The **Bektau-ata Massif**, 50 kilometers north of the city of Balkhash (Qaraghandy) in central Kazakhstan is a beautiful granite dome, which was perfectly shaped by erosion and rises about 1,000 meters above the surrounding steppe. On a clear day, it can be seen from as far as 100 kilometers away. The pegmatites here were mined for optical-grade fluorite in the 1950s and 1960s. They are famous for the world's best davidite-(La) twins, which can reach 8 centimeters. Prismatic apatite crystals from the pockets reach about the same size. They are green or yellow with no visible color change. Unlike the crystals from many other localities, Bektau-ata apatite crystals are typically terminated by a number of dipyramids, showing only small pinacoid faces, or none at all.

Dark-blue or purple crystals of manganese-rich apatite up to 9 centimeters were found in **Karmen-Kuus**, **Ak-Kezen'**, **Krasniy Kordon**, and many of the other rare element (LCT) pegmatites in the Kalba Mountain Range (Qalba Zhotasy) in East Kazakhstan. The crystals are always found completely encased in rock, although small (up to 1.5 centimeter), loose, pink, tabular apatite crystals occur in clay where the pegmatite contacts with the host rocks in the Ognevka-Bakennoe pegmatite field in the Kalba Range.

A few rare plates as large as 20-by-20 centimeters covered with exceptional, sky-blue, short prismatic apatite crystals up to 2 centimeters long and in association with black fluorite and sometimes with white bertrandite have been found in greisens of the **Akchatau** (Aqshatau) tungsten-molybdenum deposit (Qaraghandy). Discovered in 1936, the deposit was mined from 1941 until 1990. Pink-green crystals up to 4 centimeters were found in the similar Kara-oba deposit at **Dzhambul** village in central Kazakhstan (Qaraghandy).

Finally, in Tajikistan blue apatite is found in the **Karasu** LCT pegmatites in the Turkestan Mountains, and gray-green crystals up to 4 centimeters occur in an oligoclase pegmatite in the **Kukhi-Lal** spinel deposit in the western Pamir Mountains. And small flattened apatite crystals are found in the **Mika** pegmatite, in the Rangkul' area in the eastern Pamirs.

ACKNOWLEDGMENTS: *The author wishes thank Maria Alferova for her invaluable assistance with this article.*

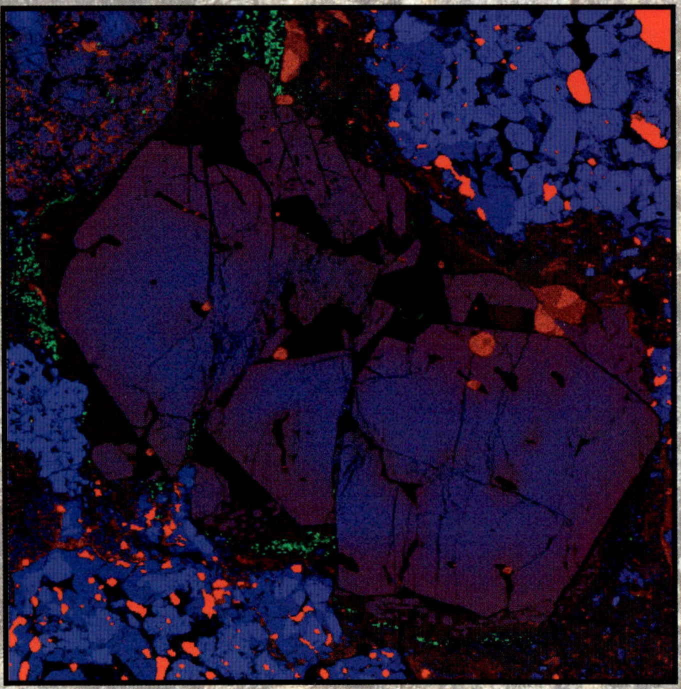

Meteoritic Apatite

by Rhian Jones, *University of New Mexico, Albuquerque*

Meteorites are wanderers from space that have crossed Earth's path, bringing with them a geological record from places far away and times long ago. Most meteorites originate from asteroids, and some come from the Moon and Mars; thus this exotic family of rocks can potentially provide insight into the geology of a number of different rocky bodies from the inner solar system. Indeed, much of what we know about the birth and early history of the solar system comes from *Meteoritics*—the science of meteorites.

Tens of thousands of individual meteorites are recognized. These include witnessed falls as well as meteorites that have been recovered some time after falling to Earth (the latter are known as "finds"). The number of known meteorites has increased enormously in the last 30 years following sustained collection programs in Antarctica and northern Africa. The discovery of numerous falls in the Sahara Desert have resulted in the widespread availability of Lunar and Martian material as well as of many new and unusual groups of asteroidal meteorites. Meteorites are classified based on their sources and/or conditions of formation.

Many meteorites contain the phosphate minerals apatite and merrillite. Although both occur only as minor or accessory phases in their host meteorites, their unique chemical properties are useful in determining the conditions under which the host rock might have formed.

Apatite contains volatile elements including the halogens fluorine and chlorine, and hydroxyl ions, the last of which indicate that water was present when the apatite formed. Understanding the conditions of formation of apatite in meteorites aids in determining factors such as the potential sources of water on Earth and the compositions of early atmospheres on other terrestrial planets.

CHONDRITES

Deep within a molecular cloud in our Milky Way and some 4.6 billion years ago, a dense region of gas began to rotate and collapse. A new star (our Sun) formed at the center of these swirling gases, and the remainder of the solar system formed from a flattened disk of gas that circled around the Sun.

As the gas in this solar nebula disk cooled, solid mineral grains condensed, much like snowflakes condense from clouds in the Earth's atmosphere. Clumps of condensed mineral grains melted and cooled to form millimeter-sized rocky beads known as "chondrules." These ancient particles are preserved today in chondrite meteorites as are sub-micrometer grains of dust, which are collectively termed "matrix." Through sticking and accretion, the chondrites developed into increasingly larger objects—the stuff from which planets are made. We base our knowledge of the age of the solar system on dating of refractory inclusions, which are the oldest known solid particles and are found in chondrites. The mineralogy of refractory inclusions is exotic and includes a suite of calcium- and aluminum-rich minerals such as perovskite, hibonite, melilite, and grossite that are rare on Earth.

As the solar nebula cooled, the common elements magnesium, silicon, and iron condensed; and the dust became dominated by silicate minerals such as olivine and pyroxene

Right: Apatite grain (gray) in the Zag ordinary chondrite
The colored grains are olivine and pyroxene; the black (opaque) mineral is iron-nickel metal.
The specimen was photographed under a petrographic microscope with crossed polarizers.
The field of view is 0.5 mm across.
Rhian Jones photo

Facing page, left: A chondrule in a carbonaceous chondrite from Antarctica EET92105 that has been altered by the action of water. The image shows the distribution of iron (red), magnesium (blue), and phosphorus (green). The chondrule consists of large, purple, zoned olivine grains, and interstitial glassy material, which appears dark. Phosphorus from the interior of the chondrule has been mobilized by water and precipitated as tiny merrillite grains (green) around the exterior of the chondrule.
The field of view is 1 mm across.
Paul Burger photo

Facing page, right: Tiny, rocky chondrules make up this slice of Northwest Africa 5205, a chondrite meteorite found in the Sahara Desert in 2006. The individual chondrules are from 1-4 mm across. Bob King collection and photo

as well as metallic iron-nickel alloys and the iron-sulfide mineral troilite. Much of the phosphorus, which would later be incorporated into apatite and merrillite, condensed within the alloys, which are dominantly the minerals kamacite and taenite. At the lowest temperatures, organic material as well as carbon dioxide and methane ices condensed, although only small amounts of water ice condensed in the inner solar system, where the terrestrial planets now reside.

Within a couple of million years after the first solids appeared in the solar nebula, the individual particles of chondrites accreted to form asteroids as much as tens of kilometers across. In the most pristine chondrites, the original properties of the 4.6-billion-year-old chondrules and matrix have been preserved. However, most chondrites experienced some geological processing on asteroids, at which point grains of apatite and merrillite first made their appearance.

Within a few tens of millions of years of the formation of the solar system, heat produced by collisions between asteroids as well as by the decay of short-lived radionuclides, such as ^{26}Al (inherited from the pre-solar molecular cloud), had brought geologic changes to the asteroids. With mild heating (to about 200 degrees Celsius), the water ice on chondritic asteroids melted, and the ensuing reactions produced hydrated minerals such as serpentine. Such "aqueous alteration" is most commonly observed in carbonaceous chondrites, which contain a relatively high abundance of fine-grained matrix and a few weight percent carbon. Dating of secondary carbonate minerals indicates that the aqueous alteration episodes occurred within a few million years of the formation of the first solar system solids.

Under more intense heat (up to about 950 degrees Celsius), the chondritic asteroids underwent thermal metamorphism: the original chondrite material recrystallized, mineral compositions became homogenized, and secondary phosphate minerals developed. Grains up to about 0.5 millimeters of both apatite and merrillite are ubiquitous in metamorphosed chondrites. In ordinary chondrites—the most common type of meteorite—apatite and merrillite together account for about one percent of the volume of the rock.

At low metamorphic temperatures, phosphorus from metal grains reacted to form merrillite. The apatite, in this case chlorapatite, occurs as singular, well-formed grains or occasionally in veins. Chlorapatite also formed as a replacement product of merrillite. The textures and compositions of the apatite grains indicate that halogen-rich fluids were present on the parent asteroids.

Apatite hosts important trace elements, including uranium and thorium which are long-lived radionuclides that can be used to date the formation of the apatite. Some of the oldest known apatite in the solar system formed 4,555 million years ago and is found in the metamorphosed ordinary chondrite, which fell in 1966 in Saint Séverin, France.

DIFFERENTIATED METEORITES

This second important class of meteorites comes from asteroids, which were heated in the first tens of millions of years after accretion such that they were extensively or even completely melted. The dense molten iron components of these asteroids sank to the cores of the bodies, and the asteroids became "differentiated," or separated, into iron cores and rocky mantles. Meteorites derived from such asteroids include "irons," which are composed predominantly of iron-nickel metal and were derived from the cores of the asteroids; "stony irons," which are mixtures of metal and silicate minerals; and "achondrites," which are derived from the asteroids' mantles.

Achondrites are comprised of silicate rocks that have an igneous (molten) history: they formed as lava flows and intrusions on their asteroid parent bodies. They are mostly

Upper left: Polished slice of the iron "Taza" NWA 859 found in 2000, Morocco
Matteo Chinellato collection and photo

Middle left: Polished slice of the stony iron Imilac meteorite found in 1822, Atacama Desert, Chile
Matteo Chinellato collection and photo

Lower left:Polished slice of the achondrite eucrite meteorite Palo Blanco Creek found in 1954, New Mexico, U.S.A.
University of New Mexico collection, Geoff Notkin photo

Below: Back-scattered electron image of fluorapatite in the eucrite (basaltic meteorite) Millbillillie, thought to have originated on asteroid 4 Vesta
The apatite grain is approximately 10 micrometers (0.01 mm) across. Rhian Jones photo

basaltic, and their mineralogy is dominated by pyroxene and plagioclase. Achondritic apatite and merrillite are igneous, having formed from the pockets of melt that remained after most of the rock had crystallized. They occur as small accessory grains, typically less than 20 micrometers across. The image on the right shows an apatite grain in an achondrite from the "HED" (Howardite, Eucrite, and Diogenite) group. The host meteorite is thought to have come from asteroid Vesta, which was visited recently by NASA's Dawn mission. Like most achondrite apatite, the HED grains are fluorapatite, which is consistent with an igneous origin. The fluorapatite is also low in OH content indicating that the magma from which it crystallized was very dry.

LUNAR AND MARTIAN METEORITES

Meteorites from asteroids record processes that took place in the first tens of millions of years of our solar system's history and on relatively small bodies (only three asteroids are larger than 500 kilometers in diameter). We are thus fortunate to have two additional classes of meteorite, namely Lunar and Martian that derived from much larger bodies: the Moon and Mars, respectively. These meteorites, as well as samples returned by missions to the Moon, allow us to examine and compare the complex geologies of planet-sized objects in our inner solar system.

There are more than 80 different known Lunar meteorites. We are certain that they come from the Moon because we can directly correlate their chemistries and mineralogies to Lunar samples collected by the Apollo and Luna programs. Lunar meteorites are ejected from the Moon when an object such as a large asteroid or comet impacts its surface.

From Lunar samples, we know that the Moon formed following the giant collision of a Mars-sized object named "Theia" with the Earth about 60 million years after the solar system began to form. After the collision, the Moon was almost entirely molten, and its geological history is dominated by the subsequent crystallization of this Lunar magma ocean. Fluorapatite and merrillite both occur in Lunar rocks as small accessory mineral grains that grew in the final stages of crystallization of basaltic magmas (between about 4 and 3 billion years ago) (photo facing page, middle left). Some basalts from the Lunar mare regions (the dark areas visible on the Moon) are enriched in phosphorus (P) as well as potassium (K) and rare-earth elements (REE). The "KREEP" component of Lunar magmas represents melt dregs from crystallization of the Lunar magma ocean. Merrillite in Lunar basalts can contain very high amounts, up to 12 weight percent, REE oxides, and apatite can contain up to 1 weight percent REE oxides.

Analyses of apatite in Lunar basalts have provided insight about the water content of the Moon's interior. Recent studies of the OH contents of Lunar apatite indicate that the Moon is not as dry as had been once assumed. The presence of a small amount of water in the Moon's mantle puts constraints on models for the Moon's formation as well as the conditions under which melting occurred to produce Lunar mare magmas.

There has been intensive study of Martian meteorites in the last 30 years, since it was first suggested that a notable class of meteorites could have originated on the Red Planet. Today,

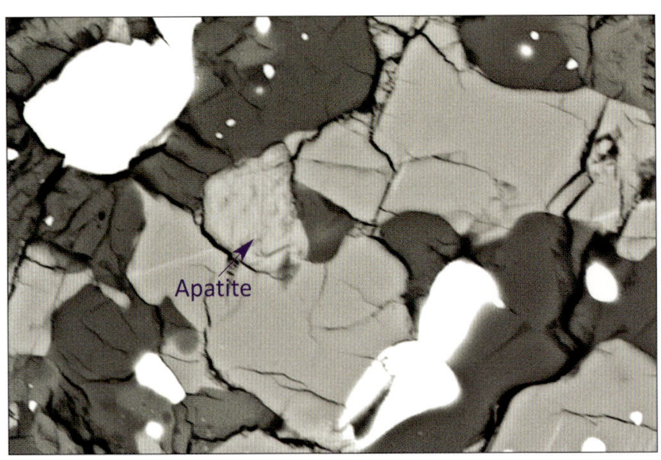

Right: Lunar meteorite Allan Hills 81005 found in Antarctica in 1981 or 1982, NASA collection and photo

Top left: Petrographic microscope image of a high-titanium basalt collected by the Apollo 17 mission. The colored grains are pyroxene, gray grains are plagioclase and the opaque (black) mineral is ilmenite. Image was taken with crossed polarizers and is 4 mm across. Rhian Jones photo

Middle left: Apatite in Apollo 15 mare basalt sample no. 15058 This image, taken with time-of-flight secondary ion mass spectrometry, shows the distribution of fluorine, oxygen and hydrogen. Fluorapatite, which shows as green, is associated with feldspar (red) and pyroxene (dark areas). The field of view is 0.2 mm across. Francis McCubbin photo

Bottom left: Back-scattered electron image of Martian meteorite QUE 94201, showing many dark gray, needle-like grains of apatite. The light gray is olivine, black is a mixture of maskelynite (highly shocked feldspar) and silica, and white is pyrrhotite. The field of view is 0.2 mm across. Francis McCubbin photo

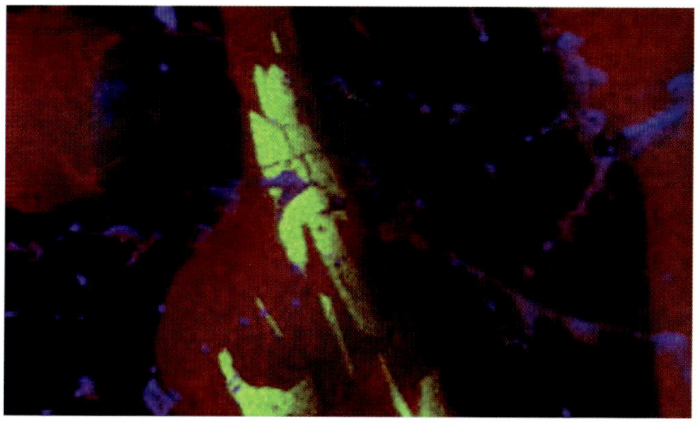

more than 60 different Martian meteorites are recognized. Unlike Lunar meteorites, there is no conclusive proof that Martian meteorites come from Mars—this must wait until we return a sample from the planet—but a key to linking these rocks to Mars is their age: they are igneous rocks that crystallized as recently as 180 million years ago thus arguing the point that they must be derived from a large planetary body. The most compelling evidence that these samples originated on Mars is that they contain pockets of glass that formed during the impact which ejected them from the planet. The composition of the gasses trapped in this glass very closely matches the composition of the Martian atmosphere.

Martian meteorites are composed of basalts and related intrusive igneous rocks, including cumulate pyroxenites and dunites (olivine cumulates). Similar to Lunar and asteroidal basalts, apatite is present as a late-crystallizing accessory mineral, with grain sizes typically less than about 50 micrometers (photo, bottom left). Analyses of apatite in Martian rocks can shed light on the abundance of volatile elements in the Martian interior. This is of fundamental interest because it enhances our knowledge of the amount of water present on Mars, both today and in the past. And an understanding of the planet's water supply is integral to our assessing the potential for the origin and development of life on the planet. Apatite in Martian meteorites has variable chlorine/fluorine ratios and contains about 0.5 weight percent water. The OH content of apatite demonstrates that the Martian mantle contains an amount of water that is comparable to that of the Earth's mantle.

Grains of extraterrestrial apatite may not compare aesthetically with some of the large, colorful and well-formed apatite crystals that formed on Earth. But the scientific importance of these tiny grains more than compensates for their modest appearance. The great age and exotic origins of phosphate minerals in meteorites as well as the clues that apatite contains about the volatile inventories of a variety of different rocky bodies contribute significantly to our understanding of the formation and geological evolution of the inner solar system.

Fluorapatite and quartz
6 cm wide
Wāna
South Wazīristān Agency
Federally Administered
Tribal Areas (FATA)
Pakistan
Steve Smale collection
Jeff Scovil photo

Literature and Cited Works

Aguilera-Serrano, J.G. 1898. "Catálogo Sistemático y Geográfico de las Especies Mineralógicas de la República Mexicana." *Boletín de la Instituto de Geología* 11:158.

Akizuki, M. 1994. "Sector Growth and Symmetry of (F, OH) Apatite from the Ashio Mine, Japan." *Mineralogical Magazine* 58:307-314.

Azevedo, R.P., and M. Calvo. 1997. "Panasqueira. Mineralogía." *Bocamina Num. Esp.* 12-27.

Azevedo, R.P., and B. Sáinz de Baranda. 1997. "Panasqueira. Historia." *Bocamina Num. Esp.* 34-41.

Azevedo, R.P., and B. Sáinz de Baranda. 1997. "Panasqueira. Geología." *Bocamina Num. Esp.* 6-11.

Bakhtin. A.I., N.M. Nizamutdinov, and V.M. Vinokurov. 1972. "Optical Absorption Spectra and Color of Apatites" (in Russian). *Seriya Geologicheskaya* 12:102-108. Moscow: Izvestiya Akademii Nauk SSSR.

Bancroft, P. 1984. *Gem and Crystal Treasures*. 348-350. Fallbrook (CA):Western Enterprises.

Bandy, M.C. 1976. "Mineralogy of Llallagua, Bolivia." Tucson Gem and Mineral Society *Special Paper #1*.

Bariand, P., and Poullen, J.F. 1978. "Famous Mineral Localities: The Pegmatites of Laghman-Nuristan-Afghanistan." *The Mineralogical Record* 9:301-308.

Barton, M.D., and D.A. Johnson. 1996. "An Evaporitic-Source Model for Igneous-Related Fe-Oxide (REE-Cu-Au-U) Mineralization." *Geology* 24:259-262.

Barton, M.D., and D.A. Johnson. "Alternative Brine Sources for Fe-Oxide (Cu-Au) Systems: Implications for Hydrothermal Alteration and Metals." In: *Hydrothermal Iron Oxide Copper-Gold and Related Deposits a Global Perspective*, edited by M.T. Porter. 43-60. Glenside (South Australia): T.M. Australian Mineral Foundation, 2000.

Bermanec, V., R. Scholz, J. Karfunkel, L. Palinkaš, D. Barišić, and G. Kniewald. 2002. *Solid Inclusions of Th-Minerals in Blue Fluorapatite from Ipirá Pegmatite, Bahia, Brazil.* Paper presented at the 11th Quadrennial IAGOD Symposium and Geocongress in Windhoek, Namibia, 22nd-26th July 2002).

Bevan, A., and J. de Laeter. 2002. *Meteorites: A Journey through Space and Time.* Washington, DC: Smithsonian Books.

Birkinbine, J.L.W. 1885. "The Cerro de Mercado (Iron Mountain) at Durango, Mexico." American Institute of Mining Engineers *Transactions* 13:189-209.

Blasse, G., and B.C. Grabmaier. 1994. *Luminescent Materials*. Berlin: Springer-Verlag.

Bloot, C., and L.C.M. de Wolf. 1953. "Geological Features of the Panasqueira Tin-Tungsten Ore-Occurrence (Portugal)." *Bol. Soc. Geol. Portugal* 11:1-58.

Bowersox, G.W., and B.E. Chamberlin. 1995. *Gemstones of Afghanistan*. Prescott (AZ): Geoscience Press.

Brauns, R. 1912. *The Mineral Kingdom*. Translated with additions by L.J. Spencer. Philadelphia: J. B. Lippincott Company.

Brezina, A. 1871. "Die Sulzbacher Epidote im Weiner Museum." *Mineralogische Mitteilungen* 1:49-52.

Broman, C., J.O. Nyström, F. Henríquez, and M. Elfman. 1999. "Fluid Inclusions in Magnetite—Apatite Ore from a Cooling System at El Laco, Chile." *GFF* 121:253-267.

Burkart, H.J. 1858. "Ueber die Fundorte der Mexikanischen Meteoreisen-Massen, als Nachtrag zu den Frueheren Angaben ueber diesen Gegenstand, unter Anschluss eines Berichts von Freidr. G. Weidner ueber das Magneteisenstein-Vorkommen an dem Cerro de Mercado bei Durango." *Neues Jahrbuch fur Mineralogie, Geologie und Palaeontologie* 1858:769-800.

Burke, E.A.J. 2008. "Tidying up Mineral Names: an IMA-CNMNC Scheme for Suffixes, Hyphens and Diacritical Marks." *The Mineralogical Record* 39:131-135.

Bussink, R.W. 1984. "Geochemistry of the Panasqueira Tungsten-Tin Deposit, Portugal (Tesis)." *Geologica Ultraiectina* 33:170.

Cantelar, E., G. Lifante, T. Calderón, R. Meléndrez, A. Millán, M.A. Àlvarez, and M. Barboza-Flores. 2001. "Optical Characterization of Rare Earths in Natural Fluorapatite." *Journal of Alloys and Compounds* 323:851-854.

Cardoso d'Orey, F. 1971. "Tungsten-Tin Mineralization and Paragenesis in the Panasqueira and Vale da Ermida Mining Districts, Portugal." *Com. Serv. Geol. Portugal* 52:116-167.

Carnot, A. 1896. "Sur les Variations Observées dans la Composition des Apatites." *Compt. Rend. Acad. Sci.* 122:1375-1380.

Cassedanne, J.P. 1986. "The Urucum Pegmatite, Minas Gerais, Brazil." *The Mineralogical Record* 17(5):307-314.

Cassedanne, J.P., and J. Alves. 1990. "Apatite and Muscovite from the Zé Pinto Prospect, Minas Gerais, Brazil." *The Mineralogical Record* 21:405-408.

Cassedanne, J.P., J.O. Cassedanne, and D.A. Sauer. 1980. "The Cruzeiro Mine (Brazil), Past and Present." *The Mineralogical Record* 11(6):363-367.

Černý, P., and T.S. Ercit. 2005. "The Classification of Granitic Pegmatites Revisited." *The Canadian Mineralogist* 43:2005-2026.

Chamberlain, S., G. Robinson, and P. Richards. 1989. "Mineralogy of the Alpine Veins Near Sherbrooke, Quebec." *The Mineralogical Record* 20(3):209-220.

Charpentier, J.F.W. 1778. *Mineralogische Geographie der Chursächsischen Lande.* Leipzig: s.n. Crusius.

Chishitsu, C. 1902. *Outlines of the Geology of Japan: Descriptive Text to Accompany the Geological Map of the Empire.* Tokyo: Imperial Geological Survey of Japan.

Chrustschoff, K. von. 1878. *Einiges ueber den Cerro del Mercado bei Durango in Mexiko.* Würzburg: G. Otto Buchdruckerei.

Chukhrov, F.V., editor. 1978. *Mineralogy of Khibiny Massif* (in Russian) 1,2:233-591. Moscow: Nauka.

Cohen, W.E. 1893. "Cerro de Mercado, Mexiko." *Zeitschrift der Kristalografie* 3:632.

Conde, L.N., V. Pereira, A. Ribeiro, and D. Thadeu. 1971. "Jazigos Hipogénicos de Estanho e Volframio." *I Congreso Hispano Luso Americano de Geología Económica.* Lisboa:Direcçâo Geral de Minas e Serviços Geologicos 19-25.

Cornejo, C., and A. Bartorelli. 2010. *Minerals and Precious Stones of Brazil.* São Paulo, Brazil: Solaris Cultural Publications.

Cotelo, J.M. 1944. "Jazigos Portugueses de Casiterite e de Wolframite." *Com. Serv. Geol. Portugal* 25:1-251.

Damour, A.A. 1856. "Sur l'Hydro-Apatite, Espèce Minérale." *Ann. Mines* 10:65-68.

Dana, E.S. 1892. *System of Mineralogy.* New York: John Wiley and Sons, Inc.

Dardenne, K., D. Vivien, and D. Huguenin. 1999. "Color of Mn(V)-Substituted Apatites $A_{10}((B,Mn)O_4)_6F_2$, A=Ba, Sr, Ca; B=P, V." *Journal of Solid State Chemistry* 146:464-472.

De Ascensao Guedes, R. 2002. "Le Coteau Minier de Panasqueira, Beira Baixa, Portugal." *Le Règne Minéral* 43:1-2, 6-32.

De Escudero, J. A. 1849. *Noticias Estadísticas del Estado de Durango, Reunidas, Aumentadas y Presentadas a la Comisión Estadística Militar.* Self-published.

De Landero, C.F. 1888. *Sinopsis Mineralógica o Catálogo Descriptivo de los Minerales.* Mexico City: La Secretaria de Fomento.

Del Rio, D.A.M. 1832. *Elementos de Orictognosia.* Mexico City: Don Mariano Joseph de Zúñiga y Ontiveros.

Dorozhkin, S.V. 2008. "Calcium Orthophosphate Cements for Biomedical Application." *Journal of Materials Science* 43:3028-3057.

Downs, R.T. 2006. "The RRUFF Project: An Integrated Study of the Chemistry, Crystallography, Raman and Infrared Spectroscopy of Minerals." Program and Abstracts of the 19th General Meeting of the International Mineralogical Association in Kobe, Japan.

Dunn, P.J. 1977. "Apatite, A Guide to Species Nomenclature." *The Mineralogical Record* 8(2):78-82.

Dunn, P.J. 1995. *Franklin and Sterling Hill, New Jersey: The World's Most Magnificent Mineral Deposits.* 5:649-650. Franklin (NJ): Franklin-Ogdensburg Mineralogical Society.

Elliott, J.C. 1994. *Structure and Chemistry of the Apatites and Other Calcium Orthophosphates.* Amsterdam: Elsevier.

Ellsworth, H.V. 1932. "Rare-Element Minerals of Canada." Geological Survey of Canada *Economic Geology Series* 11.

Emslie, R.F. 1978. "Anorthosite Massifs, Rapakivi Granites, and Late Proterozoic Rifting of North America." *Precambrian Research* 7:61-98.

Farrington, O.C. 1904. "Observations on the Geology and Geography of Western Mexico, Including an Account of the Cerro Mercado." Field Columbian Museum *Geological Series* 2:197-228.

Filippelli, G.M. 2011. "Phosphate Rock Formation and Marine Phosphorus Geochemistry: the Deep Time Perspective." *Chemosphere* 84(6):759-766.

Finkelman, R.B. 1974. "A Guide to the Identification of Minerals in Geodes from Chihuahua, Mexico." *Lapidary Journal* 27:1741-1744.

Finkelman, R.B., J.J. Matzko, Ching Chang Woo, J.S. White Jr., and W.R. Brown. 1972. "A Scanning Electron Microscopy Study of Minerals in Geodes from Chihuahua, Mexico." *The Mineralogical Record* 3:205-212.

Flores, T. 1950. "Geologic and Structural Environment of the Iron Ore Deposits of Mexico." *Economic Geology* 45:105-126.

Foshag, W.F. 1929. "Mineralogy and Geology of Cerro Mercado, Durango, Mexico." Proceedings of the United States National Museum 74:23:1-27.

Fouts, C. 1995. *1994 Mineral Asset Development Report.* Bancroft: Bancroft District Chamber of Commerce.

Foxford, K.A., R. Nicholson, D.A. Polya. 1991. "Textural Evolution of W-Cu-Sn Bearing Hydrothermal Veins at Minas da Panasqueira, Portugal." *Mineralogical Magazine* 55:435-445.

Frejes, F. 1839. "Historia Breve de la Conquista de los Estados Independientes del Imperio Mexicano." *Zacatecas, Mexico* 4:198-201.

Gaft, M.G., R. Reisfeld, and G. Panczer. 2005. *Luminescence Spectroscopy of Minerals and Materials.* Berlin: Springer-Verlag.

Gaines, R.V., and S. Thadeu. 1971. "The Minerals of Panasqueira, Portugal." *The Mineralogical Record* 2(2):73-78.

Garcia, T. 1895. *Los Mineros Mexicanos.* Mexico City: La Secretaria de Fomento.

Gilinskaya, L.G., and R.I. Mashkovtsev. 1995. "Blue and Green Centers in Natural Apatites by ERS and Optical Spectroscopy data." Translated by L. Smolina. *Journal of Structural Chemistry* 36 (1): 76-86

Gleason, J.D., M.A. Marikos, M.D. Barton, and D.A. Johnson. 2000. "Neodymium Isotopic Study of Rare Earth Element Sources and Mobility in Hydrothermal Fe-Oxide(P-REE) Mineral Deposits." *Geochimica et Cosmochimica Acta* 64(6):1059-1068.

Goldschmidt, V.M. 1913. *Atlas der Kristallformen.* Heidelberg: Carl Winters.

Graeser, S. 1998. "Alpine Minerals." *Rocks & Minerals* 73:14-32.

Gramaccioli, C.M. 1978. *"Die Mineralien der Alpen."* Band I and II. Stuttgart: Stuttgart Franckh.

Grisafe, D.A., and F.A. Hummel. 1970. "Pentavalent Ion Substitutions in the Apatite Structure Part B. Color." *Journal of Solid State Chemistry* 2:167-175.

Halse, E. 1907. "The Iron Deposit of Cerro de Mercado, Durango, Mexico: An Historical Sketch." *Mining Journal of London* 82:310

Harada, Z. 1938. "Beiträge zur Kenntnis der Optischen und Chemischen Eigenschaften des Apatits von Kurokura, Kanagawa Präfektur." Journal of the Faculty of Science, Hokkaido Imperial University. Series 4, *Geology and Mineralogy* 4(1-2):11-16

Henderson, S.T., and A.M. Marsden. 1972. *Lamps and Lighting.* London: Edward Arnold Press.

Henríquez, F., and R.F. Martin. 1978. "Crystal-Growth Textures in Magnetite Flows and Feeder Dykes, El Laco, Chile." *The Canadian Mineralogist* 16:581-589.

Hewitt, D. 1967. "Phosphate in Ontario." *Ontario Department of Mines Mineral Resources Circular No.6.*

Hintze, C. 1933. *Handbuch der Mineralogie* I(4):513-514.

Hochleitner, R., and S. Weiss. 2005. "Steckbrief Apatit (Reihe). Die Komplette

ELBAITE WITH LEPIDOLITE, 7.7 CM TALL; JONAS MINE, CONSELHEIRO PENA, MINAS GERAIS, BRAZIL; JEFF SCOVIL PHOTO

Information über die Calcium-Phosphate." *Lapis* 30(12):8-13.

Hoskyns-Abrahall, J.L. "Über die Zusammensetzung des Apatits." PhD thesis, University of Munich, 1889.

Hounslow, A.W., Chao, G.Y. 1970. "Monoclinic Chlorapatite from Ontario." *The Canadian Mineralogist* 10:252-259.

Hughes, J.M., and J. Rakovan. 2002. "The Crystal Structure of Apatite: $Ca_5(PO_4)_3$(F,OH,Cl) (in Phosphates; Geochemical, Geobiological and Materials Importance)" *Reviews in Mineralogy and Geochemistry*. 48:1-12.

Hughes, J.M., A. Ertl, H.J. Bernhardt, G.R. Rossman, and J. Rakovan. 2004. "Mn-Rich Fluorapatite from Austria: Crystal Structure, Chemical Analysis, and Spectroscopic Investigations." *American Mineralogist* 89:629-632.

Hyršl, J., and A. Petrov. 2006. "Famous Mineral Localities: Llallagua, Bolivia." *The Mineralogical Record* 37:117-162.

Hyršl, J., J. Crowley, R.H. Currier, and T. Szenics. 2011. *Peru: Paradise of Minerals*. Lima: Museo de Minerales Andrés del Castillo.

Iron Mountain Company. 1882. *The Iron Mountain Company of Durango, Mexico*. Self-published

Isaacs, A.M., and D.R. Peacor. 1981. "Panasqueiraite, a New Mineral: the OH-Equivalent of Isokite." *The Canadian Mineralogist* 9:89-392.

Isaacs, A.M., and D.R. Peacor. 1982. "The Crystal Structure of Thadeuite." *American Mineralogist* 67:120-125.

Isaacs, A.M., D.R. Peacor, and W.C. Kelly. 1979. "Thadeuite, a New Mineral from Panasquira, Portugal." *American Mineralogist* 64:359-361.

Ivanyuk, G.Yu., and V.N. Yakovenchuk. 1997. *Minerals of the Kovdor Massif.*

Jahns, R.H. 1982. "Internal Evolution of Pegmatite Bodies." In: *Granitic Pegmatites in Science and Industry,* edited by P. Černý. Mineralogical Association of Canada Short Course Handbook 8:293-327.

Johnson, P.D., J.S. Prener, and J.D. Kingsley. 1963. "Apatite: Origin of Blue Color." *Science* 141:1179-1180.

Johnson, P.W. 1963. "Cerro de Mercado Mine, Durango, Mexico." *The Mineralogist* 31(1):4-6.

Johnson, P.W. 1965. *A Field Guide to the Gems and Minerals of Mexico*. Mentone (CA): Gembooks.

Joyce, D.K. 2006. Calcite Vein-Dikes of the Grenville Geological Province. *Rocks & Minerals* 81:34-42.

Jurgeit, M. 2001. "Fluorapatit und sein Begleitmineralien von Cerro Del Mercado in Durango, Mexiko." *Mineralien Welt* 12(4):56-61.

Kalinin, P.V., and B.M. Ronenson. 1957. "Geology-Structural Features and Genesis of Slyudyanka Phlogopite Deposits." *Sovetskaya Geologiya* 58:56-73.

Kantor, B.Z. 2013. "My Favorite Mineral Specimens from Dashkesan." *Mineralogical Almanac* 18(1):28-35.

Kay, M.I., R.A. Young, and A.S. Posner. 1964. "Crystal Structure of Hydroxyapatite." *Nature* 204:1050-1052.

Kelly, W.C., and R.O. Rye. 1979. "Geologic, Fluid Inclusions and Stable Isotope Studies on the Tin-Tungsten Deposits of Panasqueira." *Portugal Economic Geology* 74:1721-1822.

King, V.T., and E.E. Foord. 1994. *Mineralogy of Maine. Volume 1: Descriptive mineralogy.* 418. Augusta: Maine Geological Survey.

Klaproth, M.H. 1788. "Phosphorsäure, ein Bestandtheil des Apatits." *Bergmännisches Journal* 1:294-300.

Klopotov, K.I. 2006. "Slyudyanka: a Brief Mineralogical Guide." *Mineralogical Almanac* 10:6-30.

Knutson, C., D.R. Peacor, and W.C. Kelly. 1985. "Luminescence, Color and Fission Track Zoning in Apatite Crystals of the Panasqueira Tin-Tungsten Deposit, Beira-Baixa, Portugal." *American Mineralogist* 70:829-837.

Kogarko, L.N., V.A. Kononova, M.P. Orlova, and A.R. Woolley. 1995. *The Alkaline Rocks and Carbonatites of the World. Part II: Former USSR*. 226. London: Chapman and Hall.

Kreibich, M., and T. Guettler. 2012. "Die Zinngrube Ehrenfriedersdorf." *Lapis* 37(7-8):22-28.

Labarthe-Hernández, G., M. Tristán-González, and A. Aguillón-Robles. 1987. "Análisis de la Mina Cerro de Mercado, Durango." *Geomimet* 150:6-14.

Lagerwey, A.A.F. 1977. "Detection of Trace Elements in Apatite Crystals from Panasqueira, Portugal, by Non-Destructive Optical Methods, Especially Polarospectrography." *Scripta Geologica* 42:1-52.

Landes, K.K. 1925. "The Paragenesis of the Granite Pegmatites of Central Maine." *American Mineralogist* 10:355-411.

Leybov, M.B. 2013. "Dashkesan: Notes on Geology and Mineralogy." *Mineralogical Almanac* 18(1):6 27.

Liu, G. 2006. *Fine Minerals of China*. Switzerland: AAA Minerals AG.

London, D. 2008. "Pegmatites." The Canadian Mineralogist, Special Publication 10.

Long, J.M., J. Rakovan, J. Jaszczak, A.J. Sommer, and R. Anczkiewicz. 2013. "Fluorapatite from a Remarkable Occurrence of Graphite and Associated Minerals, Karo Pit, Block D, Merelani Hills, Arusha Region, Tanzania." *Rocks & Minerals* 88:179-183.

Lúcio, A., and R.V. Gaines. 1973. "The Minerals of Morro Velho Gold Mine, Brazil." *The Mineralogical Record* 4(5):224-229.

Luo, Y., J. Rakovan, Y. Tang, M. Lupulescu, J.M. Hughes, and Y. Pan. 2011. "Crystal Chemistry of Th in Fluorapatite." *American Mineralogist* 96:23-33.

Lyons, J.I. Jr. 1988. "Volcanogenic Iron Oxide Deposits, Cerro de Mercado and Vicinity, Durango." *Economic Geology* 83(8):1886-1906.

MacFarlane, D. 1999. "Assessment Report for Mining Claim #5174799; Lot 17, Range 5, Hudderfield Township, Pontiac County, Quebec."

Martin, R.F. 2013. "The Late-Stage Mini-Flood of Ca in Granitic Pegmatites: A Working Hypothesis." *Sixth International Symposium on Granitic Pegmatites*, PEG2013 (New Hamphire and Maine), Keynote address.

Martin, R.F., and C. De Vito. 2005. "The Patterns of Enrichment in Felsic Pegmatites Ultimately Depend on Tectonic Setting." *The Canadian Mineralogist* 43:2027-2048.

McConnel, D. 1973. *Apatite*. New York: Springer-Verlag.

McDowell, F.W., and S.E. Clabaugh. 1979. "Ignimbrites of the Sierra Madre Occidental and Their Relationship to the Tectonic History of Western Mexico" *Geological Society of America Special Paper 180* 113-124.

McLelland, J.M., B.W. Selleck, M.A. Hamilton, and M.E. Bickford. 2010. "Late-to Post-Tectonic Setting of Some Major Proterozoic Anorthosite-Mangerite-Charnockite-Granite (AMCG) Suites." *The Canadian Mineralogist* 48:729-750.

McSween Jr., H. Y. 1999. *Meteorites and Their Parent Planets* (Second Edition). Cambridge: Cambridge University Press.

Megaw, P.K.M., and M.D. Barton. 1999. "The Geology and Minerals of Cerro de Mercado, Durango, Mexico." *Rocks & Minerals* 74:20-28.

Mellaart. J. 1964. "Excavations at Catal Huyuk, 1963." *Anatolian Studies* 14:95-97.

Menezes Filho, L.A.D. 2009. "Famous Mineral Localities: The Sapo Mine, Ferruginha District, Conselheiro Pena, Minas Gerais, Brazil." *The Mineralogical Record* 40(4):272-292.

Menzies, M.A. 1995. "The Mineralogy, Geology and Occurrence of Topaz." *The Mineralogical Record* 25:5-53.

Morteani, G., Y.A. Kostitsyn, H.A. Gilg, C. Preinfalk, and T. Razakamanana. 2013. "Geochemistry of Phlogopite, Diopside, Calcite, Anhydrite and Apatite Pegmatites and Syenites of Southern Madagascar: Evidence for Crustal Silicocarbonatitic (CSC) Melt Formation in a Panafrican Collisional Tectonic Setting." *International Journal of Earth Sciences*.

Morteani, G., C. Preinfalk, and A.H. Horn. 2000. "Classification and Mineralization Potential of the Pegmatites of the Eastern Brazilian Pegmatite Province." *Mineralium Deposita* 35:638-655.

Moyd, L. 1990. "Davis Hill near Bancroft, Ontario: an Occurrence of Large Nepheline, Biotite, Albite-Antiperthite Crystals in Calcite-Cored Vein-Dikes." *The Mineralogical Record* 21:235-245.

Naeser, C.W., and R.L. Fleischer. 1975. "Age

of the Apatite at Cerro de Mercado, Mexico; a Problem for Fission-Track Annealing Corrections." *Geophysical Research Letters* 2(2):67-70.

Nambu, M. 1970. *Introduction to Japanese Minerals*. Edited by Editorial committee for "Introduction to Japanese Minerals" Organizing Committee, International Mineralogical Association–International Association on the Genesis of Ore Deposits meeting '70. Geological Survey of Japan.

Naslund, H.R., and F. Henríquez. 2004. *Field Guide to the El Laco Magnetite Deposits*. IAVCEI General Assembly (Pucón, Chile), Field Excursion Guidebook.

Nimura, K. 1998. *The Ashio Riot of 1907: A Social History of Mining in Japan*. Translated by Terry Boardman. Durham (NC): Duke University Press.

Norton, O.R. 2002. *Cambridge Encyclopedia of Meteorites*. Cambridge: Cambridge University Press.

Nyström, J.O., and F. Henríquez. 1994. "Magmatic Features of Iron Ores of the Kiruna Type in Chile and Sweden: Ore Textures and Magnetite Geochemistry." *Economic Geology* 89:820-839.

Orris, G.J., and J.D. Bliss. 2002. "Mines and Mineral Occurrences of Afghanistan." *United States Geological Survey Open-File Report* 02-110.

Ottens, B. 2005. "Xuebaoding, Pingwu County, Sichuan Province, China." *The Mineralogical Record* 36:45-57.

Ottens, B. 2008. *China, Mineralien, Fundstellen und Lagerstätten*. München: TU Bergakademie Freiberg/Weise Verlag. 552p.

Ottens, B. 2011a. *Indien, Mineralien, Fundstellen und Lagerstätten*. München: TU Bergakademie Freiberg / Weise Verlag. 384p.

Ottens, B. 2011b. "The Yaogangxian Mine, Hunan Province, China." *The Mineralogical Record* 42:557-603.

Ottens, B., and R.B. Cook. 2005. "The Yaogangxian Tungsten Mine, Yizhang County, Chenzhou, Hunan Province, China." *Rocks & Minerals* 80:46-57.

Ottens, B., and G. Neumeier. 2012. The "Huanggang Mine, Inner Mongolia, China." *The Mineralogical Record* 37:529-563.

Palache, C., H. Berman, and C. Frondel. 1951. *The System of Mineralogy of James Dwight Dana and Edward Salisbury Dana*. London: John Wiley and Sons, Inc.

Pan, Y., and M.E. Fleet. 2002. "Compositions of the Apatite Group Minerals: Substitution Mechanisms and Controlling Factors." In: *Phosphates: Geochemical, Geobiological and Materials Importance*, edited by M. Kohn, J. Rakovan, and J.M. Hughes. Reviews in Mineralogy and Geochemistry 48:13-49.

Parker, R.L. 1954. *Die Mineralfunde der Schweizer Alpen*. Basel: Wepf & Co.

Pasero, M., A.R. Kampf, C. Ferraris, I.V. Pekov, J. Rakovan, and T. White. 2010. "Nomenclature of Apatite Supergroup Minerals." *European Journal of Mineralogy* 22:163-179.

Pekov, I.V. 2000. *Lovozero Massif: History, Pegmatites, Minerals*. Moscow: Vneshtorgizdat.

Pekov, I.V., and A.S. Podlenyi. 2004. *Kukisvumchorr Deposit: Mineralogy of Alkaline Pegmatites and Hydrothermalites* 168.

Petrov, A. 1997. *Mineral News* 13(4)

Piper, W.W., C. Kravitza, and R.K. Swank. 1965. "Axially Symmetric Paramagnetic Color Centers in Fluorapatite." *Physical Review* 138:1802-1814.

Quintero, A.R.M. 1923. "Gem Apatite in Cerro de Mercado-Durango." *Boletín del Instituto Geológico de México* 44.

Rakovan, J. 2002. "Growth and Surface Structure of Apatite." In: *Phosphates: Geochemical, Geobiological and Materials Importance*, edited by M. Kohn, J. Rakovan, and J.M. Hughes. Reviews in *Mineralogy and Geochemistry*. 48:51-86.

Rakovan, J. 2003. "A Word to the Wise: Skarn." *Rocks & Minerals* 78:271.

Rakovan, J. 2004. "Word to the Wise: Growth Hillock." *Rocks & Minerals* 79:415-417.

Rakovan, J. 2007. "A Word to the Wise: Greisen." *Rocks & Minerals* 82:157-159.

Rakovan, J. 2009. "Word to the Wise - Sectoral Zoning." *Rocks & Minerals* 84:171-176.

Rakovan, J., and R.J. Reeder. 1994. "Differential Incorporation of Trace Elements and Dissymmetrization in Apatite: The Role of Surface Structure During Growth." *American Mineralogist* 79:892-903.

Rakovan, J., and R.J. Reeder. 1996. "Intracrystalline Rare Earth Element Distributions in Apatite: Surface Structural Influences on Zoning During Growth." *Geochimica et Cosmochimica Acta* 60:4435-4445.

Rakovan, J., and G. Waychunas. 1996. "Luminescence in Minerals." *The Mineralogical Record* 27:7-19.

Rakovan, J., D.K. McDaniel, and R.J. Reeder. 1997. "Use of Surface-Controlled REE Sectoral Zoning in Apatite from Llallagua, Bolivia, to Determine a Single-Crystal Sm-Nd Age." *Earth and Planetary Science Letters* 146:329-336.

Rakovan, J., R.J. Reeder, E.J. Elzinga, D.J. Cherniak, C.D. Tait, and D.E. Morris. 2002. "Structural Characterization of U(VI) in Apatite by X-ray Absorption Spectroscopy." *Environmental Science and Technology* 36:3114-3117.

Ramirez, S. 1884. *La Riqueza Minera de México*. Secretaría de Fomento 404.

Rammelsberg, C.F. 1860. *Handbuch der Mineralchemie*. 1038. Leipzig: Wilhelm Engelmann.

Rangel, M.F. 1902. "Criadero de Fierro del Cerro de Mercado de Durango." *Universidad Nacional Autónoma de México, Instituto de Geológia, Boletín* 16:144.

Reis, A.C. 1971. "As Minas da Panasqueira." *Boletín de Minas (D.G.G.M.-Lisboa)* 8:3-34.

Report of the Royal Commission on the Mineral Resources of Ontario and Measures for Their Development. Toronto: Warwick & Sons.

Ribeiro, H.B., K.J. Guedes, M.V.B. Pinheiro, S. Greulich-Weber, and K. Krambrock. 2005. "About the Blue and Green Colours in Natural Fluorapatite." *Physica Status Solidi (c)*, 2:720-723.

Richards, R.P. 1990. "The Origin of Faden Quartz." *The Mineralogical Record* 21:191-201.

Roberts, W.L., and G. Rapp Jr. 1965. "Mineralogy of the Black Hills." *South Dakota School of Mines & Technology, Rapid City, South Dakota Bulletin* 18:268.

Rose, G. 1827. "Ueber die Chemische Zusammensetzung der Apatite." In: *Annalen der Physik und Chemie* 2:185-214. Leipzig: J.A. Barth.

Rossman, G.R. 2003. "The Mineral Spectroscopy Server." http://minerals.gps.caltech.edu/files/Visible/APATITE/Index.htm Pasadena, CA.

Sabina, A. 1986. "Rocks and Minerals for the Collector Bancroft-Parry Sound Area and Southern Ontario." *Geological Survey of Canada Miscellaneous Report* 39.

Sabina, A. 1987. "Rocks and Minerals for the Collector Hull-Maniwaki, Quebec; Ottawa-Peterborough, Ontario." *Geological Survey of Canada Miscellaneous Report* 41.

Sabina, A. 1971. "Rocks and Minerals for the Collector Ottawa to North Bay, Ontario; Hull to Waltham, Quebec." *Geological Survey of Canada Paper* 70-50.

Salazar-Salinas, L., P. González, M. Santillán, A. Acevedo, and A.R. Martínez-Quintero. 1923. "El Cerro de Mercado, Durango." *Universidad Nacional Autónoma de México, Instituto de Geológia, Boletín* 44:90.

Santosh, M., and A.S. Collins. 2003. "Gemstone Mineralization in the Palghat-Cauvery Shear Zone System (Karur-Kangayam Belt), Southern India." *Gondwana Research* 6(4):911-918.

Scherba, G. N. 1970. "Greisens." *International Geology Review* 12:114-150.

Seemann, R. 1986. "Knappenwand-Untersulzbachtal-Austria." *The Mineralogical Record* 17:167-181.

Shaw, D.M., R.L. Moxham, R.H. Filby, and W.W. Lapkowsky. 1963. "The Petrology and Geochemistry of Some Grenville Skarns. II. Geochemistry." *The Canadian Mineralogist* 7:578-616.

Sherwin, D., O. Clark, and the Friends of the Tay Watershed Association. "Geology, Mineral Deposits and History of Mining in the Tay River Watershed." Accessed November, 2012. http://www.tayriver.org/documents/geology_report/geology_report.htm.

Silliman, B. Jr. 1882. "Martite of the Cerro de Mercado, or Iron Mountain, of Durango, Mex., and Certain Iron Ores of Sinaloa."

Silver, Kongsberg, Norway • Jesse La Plante photo

American Journal of Science 3(24):375-379.

Sinkankas, J. 1959. *Gemstones of North America*. 657. New York, NJ: Van Nostrand Reinhold Company.

Sinkankas, J. 1976. *Gemstones of North America*, Vol. II, 248-252. New York: Van Nostrand Reinhold Company.

Sinkankas, J. 1997. *Gemstones of North America*, Vol. III. Tucson: Geoscience Press, Inc.

Sirbescu, M.L.C., M.A. Leatherman, J.J. Student, and A.R. Beehr. 2009. "Apatite Textures and Compositions as Records of Crystallization Processes in the Animikie Red Ace Pegmatite Dike, Wisconsin, USA." *The Canadian Mineralogist* 47:725-743.

Smith, C., S. Russell, and G. Benedix. 2010. *Meteorites*. Tonawanda (NY): Firefly Books.

Tello, A. 1653. "Cronica Miscelánea y Conquista Espiritual y Temporal de la Santa Provincía de Xalisco en el Nuevo Reino de la Galicía y Nueva Vizcaya y Descubrimiénto del Nuevo México." Guadalajara: *La República Literaria*, de C.L. de Guevara y ca.

Terrones-Benitez, A. 1944. "Cerro de Mercado, Mexico's Iron Mountain." *Engineering and Mining Journal* 145(9):88-89.

Thadeu, D. 1951. "Geologia do Couto Mineiro da Panasqueira." *Com. Serv. Geol. Portugal* 32(1):4-6.

Thadeu, D. 1979. "Le Gisement Stanno-Wolframifère de Panasqueira (Portugal)." *Chronique de la Recherche Minière* 450:35-42.

Traill, R. 1972 (1970). "A Catalogue of Canadian Minerals." *Geological Survey of Canada Paper,* 69-45

Tyson, R., and H. Tyson. Tyson's Fine Minerals. "Mineral Gallery-Liscombe, Collection History." Last modified 2012. Accessed November 30, 2012. http://www.tysonsfineminerals.com/mineral-gallery/?id=44

Udd, J. 2005. *A Guide to the Mineral Deposits of Southeastern Ontario and Southwestern Quebec*. Ottawa: CJ Multi-Media, Inc.

Van Velthuizen, J. 1992. "Giant Fluorapatite Crystals: A Question of Locality." *The Mineralogical Record* 23(6): 459-463.

Van Velthuizen, J., D. Sturman, G. Robinson, and G. Ansell. 1992. "Mineralogy of the Rapid Creek and Big Fish River Area, Yukon Territory." *The Mineralogical Record* 23(4):1-47.

Vaz Leal, M. 1945. *As Minas da Panasqueira: Vida e História*. Reedición de Cámara Municipal da Covilha e da Junta de Freguesia de Aldeia de San Francisco de Assis, 2004.

Vaz, J.E. 1980. "Effects of Natural Radioactivity on the Thermoluminescence of Apatite Crystals at Cerro de Mercado, Mexico." *Modern Geology* 7(3):171-175.

Vincent, H.R., K.I. McConnell, and P.C. Perley. 1990. "Geology of Selected Mafic and Ultramafic Rocks of Georgia: a Review." Georgia Department of Natural Resources, Environmental Protection Division, Georgia Geologic Survey Information circular #82.

Voileau. A. 1975. "Panasqueira." *Monde et Minéraux* 12:286 and 13:322.

Wada, T. 1904. *Minerals of Japan*. Translated by T. Ogawa. Tokyo.

Ward, H.G. 1829. *Mexico During the Years 1825, 1826 and Part of 1827: with an Account of the Mining Companies and of the Political Events in the Republic to the Present Day.* 347. London: H. Colburn.

Waychunas, G.A. 2002. "Apatite Luminescence." In: *Phosphates: Geochemical, Geobiological and Materials Importance*, edited by M. Kohn, J. Rakovan, and J.M. Hughes. Reviews in *Mineralogy and Geochemistry*, 48:701-742.

Weidner, F. 1858. *El Cerro de Mercado de Durango.* Mexico City: Andrade y Escalante.

Weidner, F. 1878. "Informe Científico Sobre el Cerro de Mercado en Durango." *Anales del Ministerio de Fomento de la República Mexicana*, III:155-183.

Weiss, S. 2012. "Der Ehrenfriedersdorfer Sauberg-Typlokalität für Apatit." *Lapis* 37(8):42-43.

Werner, A.G. 1787. "Kurze Klassification und Beschreibung der Verschiedenen Gebirgsarten." *Abh. Bömischen Gesell. Wissen*. 1:272-297.

Werner, A.G. 1788. "Geschichte, Karakteristik, und kurze chemische Untersuchung des Apatits." *Bergmännisches Journal* 1:76-96.

Witherbee, T.F. 1902. "The Iron Mountain and the Plant of the Mexican National Iron and Steel Co., Durango, Mexico." American Institute of Mining Engineers *Transactions* 32:156-163.

Whitmore, R.W., and R.C. Lawrence. 2004. *The Pegmatite Mines Known as Palermo*. 219. Weare (NH): Palermo Mines Ltd.

Wilson, M.L., J.B. Paces, and A.P. Ruotsala. 1984. "Fluorapatite from the King Lithia Mine, Custer County, South Dakota." *The Mineralogical Record* 15:361-366.

Wilson, W.E. 2012. "Famous Mineral Localities-The Jonas Mine, Itatiaia, Minas Gerais, Brazil." *The Mineralogical Record* 43(3):289-317.

Wodzicki, W.A. "The Evolution of Laramide Igneous Rocks and Porphyry Copper Mineralization in the Cananea District, Sonora, Mexico." PhD Dissertation, University of Arizona, 1995.

Wolff, J.E., and C. Palache. 1902. "Apatite from Minot, Maine." *Proceedings of the American Academy of Arts and Science* 37:517-528.

Young, E.J., A.T. Myers, E.L. Munson, and N.M. Conklin. 1969. "Mineralogy and Geochemistry of Fluorapatite from Cerro de Mercado, Durango, Mexico." *U.S. Geological Survey Professional Paper* 650-D:D84-D93.

Zepharovich, V. von. 1869. "Neuere Mineralfundorte in Salzburg." *Jahrbuch für Geologie und Paläontologie*, XIX: 231-234.

Zharinov, A.A., V.V. Ponomarenko, and I.V. Pekov. 2008. "Color-Change Apatite From Kazakhstan." *Rocks & Minerals* 83(2):148-151.

Zubiria y Campa, L. 1922. *El Cerro de Mercado y los aerolitos*. Durango: Revista Durango.

Zubiria y Campa, L. 1924. *El Cerro de Mercado*. 79. Mexico City: Imprenta Victoria.

Fluorapatite
12.2 cm wide
Harvard Quarry
Harvard Quarry, Greenwood
Oxford County, Maine
Larry Stifler collection
Jeff Scovil photo

Rutile, quartz and hematite, 7 cm tall
Ibitiara, Bahia, Brazil
Tiziano Bonisoli collection